CONCISE PAINT TECHNOLOGY

J. BOXALL
A.M.I. Corr. T., A.T.S.C., L.I.M.F.
Higher Scientific Officer, Building Research Establishment, Princes Risborough

J. A. von FRAUNHOFER
M.Sc., Ph.D., M.I.M., F.I.Corr.T., C.Chem., F.R.I.C.
Head of Department of Dental Materials,
Institute of Dental Surgery, University of London
and Visiting Lecturer, Division of Metal Science,
Polytechnic of the South Bank, London

CHEMICAL PUBLISHING
NEW YORK

© Paul Elek (Scientific Books) Ltd., 1977

First American edition, 1977
Chemical Publishing Co., Inc.
155 W. 19 St., New York, N.Y. 10011

Printed in Great Britain by
Biddles Limited, Guildford, Surrey

To our wives and families

The more alternatives, the more
difficult the choice
            Abbé d'Allainval, 1726

Contents

Preface

1.	FUNDAMENTALS OF PAINT TECHNOLOGY	1
1.1	Classification of Paints	1
1.2	Basic Polymer Chemistry	4
1.3	General Properties of Pigments	4
1.4	Bibliography	28
2.	POLYMERS USED IN PAINTS	29
2.1	Convertible and Non-Convertible Coatings	29
2.2	Oleoresinous Binders	30
2.3	Alkyd Resins	31
2.4	Amino Resins	35
2.5	Vinyl Resins	36
2.6	Acrylic Resins	39
2.7	Epoxy Resins	40
2.8	Polyurethane Resins	44
2.9	Chlorinated Rubbers	49
2.10	Cellulosic Polymers	50
2.11	Phenolic Resins	51
2.12	Other Binder Materials	52
2.13	Bibliography	57
3.	PIGMENTATION OF PAINTS	58
3.1	Pigments, Extenders and Dyes	58
3.2	Inorganic Pigments	61
3.3	Metallic Pigments	74
3.4	Organic Pigments	75
3.5	Extenders	81
3.6	Bibliography	83

4.	ADDITIVES AND SOLVENTS FOR PAINTS	84
4.1	Plasticisers	84
4.2	Solvents	89
4.3	Drying Accelerators	97
4.4	Biocides, Fungicides and Anti-Fouling Additives	98
4.5	Pigment Dispersing Agents	102
4.6	Paint Viscosity Modifiers	104
4.7	Pigment Anti-Settling Agents	106
5.	PAINT FORMULATION	108
5.1	Principles of Formulation	108
5.2	Methods of Paint Formulation	113
5.3	Paint Manufacture	119
6.	TECHNIQUES OF PAINT APPLICATION	123
6.1	Brushes and Rollers	123
6.2	Paint Spraying	125
6.3	Dip Coating Techniques	130
6.4	Roller Coating and Curtain Coating	133
6.5	Bibliography	134
7.	SUBSTRATES	135
7.1	Metals	135
7.2	Cementitious Materials	148
7.3	Timber	153
7.4	Bibliography	161
8.	PAINT DURABILITY	162
8.1	Influence of Location on Durability	162
8.2	Influence of Weather on Durability	165
8.3	Bibliography	168
9.	TESTING OF PAINTS AND PAINT FILMS	169
9.1	The Testing of Paints	170
9.2	Properties of Wet Paint Films	173
9.3	General Properties of Dry Paint Films	177
9.4	Mechanical Properties	184
9.5	Optical Properties	190
9.6	Permeability, Weathering and Corrosion Protection	194
9.7	Bibliography	199
	Index	201

Preface

Painting is the oldest and most widespread method of achieving a decorative and, usually, protective finish for engineering and other materials. The range of substrates to which paints are applied is vast, encompassing metals, timber, plaster, cement and concrete. Furthermore, an often bewildering diversity of paints and paint types are available to users and there are a number of different methods of application.

This book has been written in response to the need for a a concise appraisal of paint technology. This appraisal is required by all paint users and those who are required to specify surface finishes for engineering structures, industrial and domestic appliances and the multiplicity of other end products requiring decorative and/or protective finishes. The very complexity of modern paints renders their selection and specification difficult and whilst authoritative, and often exhaustive, treatments of the various aspects of paint technology are available, no modern but concise and up-to-date treatment of modern paints has hitherto been generally available. This book, we feel, fills this gap in the literature of paint technology.

In order that the book should be useful to as diverse a range of readers as possible, a logical approach to the treatment has been adopted. Thus, the basic principles of polymer chemistry, pigmentation, colour and similar subjects are outlined in the first chapter. Then the major paint components, namely, the polymeric binders, pigments and additives and solvents, are each discussed in separate chapters. These are followed by chapters on the principles of paint formulation, paint application techniques and substrates for paints. Finally, there are chapters on paint durability and the testing of paints.

Since it is imposssible to provide an in-depth coverage of the whole field of paint technology in a single volume of reasonable size, a detailed but basic treatment has been adopted with recommendations for further reading being given at the end of most chapters.

J.Boxall
J.A.von Fraunhofer

1 Fundamentals of paint technology

1.1 CLASSIFICATION OF PAINTS

Paints can be defined as largely organic coatings applied to surfaces to provide both protective and decorative functions. The object of painting is to interpose a film between the substrate and its environment. This is normally achieved by the application of a paint system comprising a number of coats of paint, each one formulated to impart certain characteristics to the overall system. The performance and formulation requirements of the paint system with regard to its decorative and protective value are dependent on both the substrate and its anticipated service environment. Many paints designed to impart a high degree of protection against corrosion often have little aesthetic value, as for example highly protective bituminous paints. Conversely, many paint systems designed to provide coatings with a high degree of decorative value often can only be used on substrates exposed to relatively mild service environments. Generally, however, a compromise can be achieved between the decorative and protective functions of a coating system.

In practice, paints are applied to a very wide variety of surfaces which can differ markedly in their physical and chemical characteristics. For example, such diverse substrates as metals, concrete, plaster and woodwork are painted. Clearly these substrates may vary in their quality, that is, state of deterioration, cleanliness, surface finish, shape, accessibility, and there also may be wide differences in the severity of the service environments to which the painted surface is to be exposed.

Despite the apparent complexity of substrates that require coating, and the multiplicity of paint materials that are used to coat them, all paints are basically similar in composition in that they contain a liquid polymeric or resinous material known as a binder. It is this component which, after conversion to a solid through the paint's drying process, provides the surface film with the necessary attributes of adhesion, flexibility, toughness and durability.

Paints also contain finely dispersed solid materials called pigments. A major function of a pigment is to provide colour and opacity for the paint film, and hence for the surface to which it is applied, although certain types of pigment have other specific functions such as improving film durability or providing corrosion resistance for metal substrates. In addition, other components are usually incorporated into paints to modify the basic characteristics imposed by the pigment and resin constituents. Typical examples of such additives are extenders, which can modify the film's mechanical characteristics without imparting opacity; driers, which control the drying or curing process of the liquid paint; and fungicides, which inhibit mould growth on the film's surface during service exposure. Volatile solvents are also included in order to control the application properties of the liquid paint and modify the drying process of the film.

In practice, paints may be classified on the basis of the type of pigment (e.g. red lead paint), by the resin binder (e.g. alkyd paint), or by its proposed function or service applications (e.g. chemical resistant finish). Clearly, any single descriptive classification used in isolation has inherent disadvantages. At this stage, however, the classification of paints and allied materials by their function will be of greatest value and the characteristics of the most important classes are summarised below.

Fillers and Stoppers

These are paste-like materials, normally highly pigmented, used to fill surface imperfections (fillers) and to make good gross surface defects prior to painting operations (stoppers). A variety of non-setting materials such as mastics, sealants and putties, which are designed to stay flexible, can also be included in this class.

Sealers

These are low viscosity materials, not necessarily pigmented, whose primary function is to penetrate and seal porous surfaces such as plaster, masonry and certain timbers. Certain types may also be used to seal chemically active surfaces or to consolidate friable surfaces. Sealers do not normally deposit films of high-build although in certain situations they may be used instead of conventional primers.

Primers

Primers are the first high-build pigmented coats applied to new surfaces or to old cleaned surfaces, prior to the application of other components of the finishing system. They are invariably specifically formulated for particular substrates, but all primers regardless of other performance criteria must possess good adhesion on the substrate.

Undercoats

The undercoat is a pigmented paint applied after sealing and/or priming of a surface, but prior to the application of the finishing coats. An undercoat normally has a matt finish (so as not to impair adhesion of subsequent coats), a high pigment content (to enhance the opacity of the paint system) and a colour to complement that of the ultimate finishing coats.

Finishes

The finishing coat is the final coat of the paint system, formulated to provide both high durability and acceptable aesthetic value in the proposed service environment. Finishing coats are normally pigmented although unpigmented finishes may be used, particularly on certain metals and timber. Materials of the latter class are generally termed varnishes or lacquers, and the coating system in these instances is obtained by multicoat application of unpigmented 'finish'. The surface appearance of pigmented finishing coats can be modified to suit most requirements in terms of colour, gloss level and texture. The unpigmented finishes are normally only varied in terms of gloss level.

In practice, it is normally found that all paints (and associated products), regardless of their constituent raw materials or anticipated service requirements, can be classified into one or other of the above groups. However, the raw materials used, and their relative concentration ratios, broadly determine the performance characteristics of the desired paint film. Consequently, the primary ingredients of paints, namely the binder and the pigment, will largely determine the classification of the paint. For this reason, the basic principles of polymer technology and pigment theory will be outlined in the remainder of this chapter. In subsequent chapters the polymers, pigments and other paint ingredients will be discussed in greater detail.

1.2   BASIC POLYMER CHEMISTRY

Polymers, which are also referred to as resins or plastics, are long chain molecules formed by the joining together of large numbers of small molecules, known as monomers, or simply mers. The long molecular chains are flexible and, consequently, chain entanglement occurs readily. Furthermore, many polymers are formed from mers having bulky side groups (see below), so that the molecular chains do not pack closely together. Thus, as a result of the long and flexible chains and the loose packing upon chain entanglement, polymers are generally non-crystalline solids, although there are many exceptions. Factors contributing to non-crystallinity are the length and branching of the molecular chains, the presence of side groups (particularly bulky groups) along the chain, copolymerisation (polymers formed from two or more monomers) and plasticisation. Plasticisers are low molecular weight compounds which when added to polymers exert a softening or plasticising action on the polymer by separating the molecular chains and hence hindering crystallinity of the polymer.

Small molecules containing unsaturated groups, (e.g. carbon-carbon double bonds as in vinyl chloride) or those based on strained ring structures, can usually be polymerised, under suitable reaction conditions, to form polymers. It should be noted, however, that these polymer precursors are themselves normally stable, and polymerisation will only occur after a suitable initiation reaction.

1.2.1   Polymerisation Reactions

The formation of a polymer usually involves three processes, namely initiation, propagation and termination. A fourth process, chain transfer, can also occur.

Initiation involves attack upon the monomer molecule by an initiator such that the double bond or the strained

ring is opened and a new, reactive species is formed.
The initiator may be a free radical or an ion. A free
radical is a fragment of a larger molecule formed by
dissociation or decomposition of that molecule. A
typical initiator is dibenzoyl peroxide $C_6H_5.CO_2.O_2C.C_6H_5$
which, under the influence of heat or through reaction with
another compound (known as a promoter) such as a tertiary amine,
will decompose to form two benzoyloxy radicals $C_6H_5CO_2\cdot$.
The initiator fragment or free radical, which contains
an unpaired electron, will attack the double bond and
attach to one side of it, so that a larger radical is
formed:

$$C_6H_5-CO_2-O_2C-C_6H_5 \longrightarrow 2C_6H_5-CO_2\cdot \; \equiv \; A\cdot$$

$$A\cdot + CH_2 \!=\! CHX \longrightarrow ACH_2 \!-\! C\dot{\,}HX$$

This new free radical may then react with another
monomer molecule so that a still larger reactive species
is formed:

$$ACH_2 \!-\! \dot{C}HX + CH_2 \!=\! CHX \longrightarrow ACH_2 \!-\! CHX \!-\! CH_2 \!-\! \dot{C}HX$$

and this chain reaction can continue (propagation)
until all the free monomer is consumed or another process
intervenes (chain termination).

Addition polymerisation may also be initiated by strong
acids which, in the presence of a suitable promoter,
form ions which can attack double bonds. Typically,
boron trifluoride, a Lewis acid, will react with water
to form an ion which will react with a monomer containing
a double bond to form a carbonium ion:

$$BF_3 + H_2O \longrightarrow H^+ \; HOBF_3^-$$

$$H^+ \; HOBF_3^- + CH_2 \!=\! CHX \longrightarrow \overset{+}{C}H_2X \; F_3\bar{B}OH + \dot{C}H_2$$

and the carbonium ion can react with another monomer
molecule so that chain propagation occurs. This is known
as cationic polymerisation.

Various other initiators, such as strong bases
(anionic polymerisation) and organometallic compounds
(organometallic polymerisation), are used in polymerisation reactions, notably for vinyl compounds.

Addition polymerisation does not continue indefinitely
and, furthermore, there is a limit to the size of the
individual chains within the polymerised mass. A single

initiator molecule such as dibenzoyl peroxide or di-*tert*-butyl peroxide will, upon dissociation, form two radicals, both of which will initiate chain propagation. Addition to the chain can occur only if a monomer molecule is in the immediate vicinity of the growing chain, or if it can diffuse rapidly to the reaction site. As the number and size of the growing chains increase, there is increasing competition for residual monomer molecules. If two growing ends interact, the propagation of both chains will terminate:

$$A CH_2 \cdots CH_2\text{-}\overset{\bullet}{C}HX + \overset{\bullet}{C}HX\text{-}CH_2 \cdots CH_2 A \longrightarrow$$

$$A CH_2 \cdots CH_2\text{-}CHX\text{—}CHX\text{-}CH_2 \cdots CH_2 A$$

The length of the terminated chain will be determined by the lengths of the two individual growing chains. Chain growth can also be terminated by the addition of inhibitors, which react with the growing chains to form inactive products. Typical inhibitors are oxygen, iodine, benzoquinone and polycyclic aromatic compounds. It should also be noted that excessive addition of an initiator can have an inhibiting effect since the oxy radical can add to a growing chain and terminate propagation:

$$A CH_2 \cdots CH_2\text{-}\overset{\bullet}{C}HX + \overset{\bullet}{A} \longrightarrow A CH_2 \cdots CH_2 CHXA$$

Many monomers contain an inhibitor, known as an antioxidant, to prevent oxidation and hence premature polymerisation in storage. Consequently, when an initiator is added to the monomer there is an induction period, during which little polymerisation occurs and during which the antioxidant is destroyed, before polymerisation is initiated.

The chain length of polymer molecules may also be controlled by chain transfer agents such as carbon tetrachloride, toluene and various mercaptans. The chain transfer agent will react with a growing chain to terminate its propagation, whilst the agent itself will form a radical which can initiate the growth of a new chain:

$$A CH_2 \cdots CH_2\text{-}\overset{\bullet}{C}HX + CCl_4 \longrightarrow A CH_2 \cdots CH_2\text{-}\overset{\bullet}{C}HX\text{-}Cl + \overset{\bullet}{C}Cl_3$$

$$\overset{\bullet}{C}Cl_3 + CH_2\text{=}CHX \longrightarrow Cl_3 C\text{-}CH_2\text{-}\overset{\bullet}{C}HX$$

Chain transfer agents are used to control or regulate the chain size of polymers so that the average molecular weight is reduced, but the polymerisation rate is virtually unaltered.

Molecules containing reactive groupings, such as the carbon-carbon double bond undergo addition polymerisation. The simplest monomers contain only one double bond, for example, the monosubstituted ethylenes or vinyl compounds, $CH_2\!=\!CHX$, and the 1,1 disubstituted ethylenes or vinylidene compounds, $CH_2\!=\!CXY$. The nature of the monomers or polymer precursors and also the chain lengths or average molecular weights will determine the properties of the polymers. Thus, since a very wide variety of monomers are available, a vast range of polymers can be made. Polymers formed from a single type of monomer are known as homopolymers.

Addition polymers may also be formed from mixtures of two or more different monomers. The products are termed copolymers. Copolymerisation results from the growing chains reacting with both mers present in the reaction mix and copolymers can be made with different geometric arrangements of the individual mers in the chain. By this means the composition of copolymers can be varied and a very wide range of properties can be obtained. Common examples of mers that copolymerise are mixtures of vinyl chloride and vinyl acetate and mixtures of acrylic acids. Copolymers derived from both mixtures are important in paint technology. Copolymerisation techniques permit an even greater variety of polymers to be produced and polymers formed from two or more different types of monomer are known as heteropolymers.

An alternative system of polymerisation is condensation polymerisation in which the monomers react together, with the elimination of a small molecule, such as H_2O, HCl or CH_3OH, to form a large molecule. Typically, phenol and formaldehyde will undergo condensation polymerisation to form Bakelite, a phenolic resin:

$$H-\underset{H}{\overset{}{C}}=O \quad + \quad \text{phenol} \quad + \quad \text{phenol}$$

$$\longrightarrow \quad \text{Bakelite} \quad + \quad H_2O$$

Bakelite

Similarly, urea and other amides will condense with formaldehyde:

$$H_2N - \underset{O}{\overset{\|}{C}} - NH_2 \; + \; H - \underset{H}{\overset{|}{C}} = O \; + \; H_2N - \underset{O}{\overset{\|}{C}} - NH_2 \longrightarrow$$

$$H_2N - \underset{O}{\overset{\|}{C}} - N - H$$

$$\begin{array}{c} \diagdown H \\ C \\ H \diagup \diagdown N - \underset{O}{\overset{\|}{C}} - N - H \\ H H \end{array} \quad + \; H_2O$$

and the linear polymer formed by this reaction can undergo further condensation to form a thermosetting resin (see below).

A further reaction that can occur during polymer growth is cross-linkage. Cross-linkage is the formation of bonds between different polymer chains, the reaction involving either free radicals (as in addition polymerisation) or condensation. For example, separate chains of polybutadiene will cross-link through reaction, namely oxidation, at some of the double bonds retained in the polymer to form oxygen links between the chains:

$\sim\!CH_2-CH=CH-CH_2-\;CH_2-CH=CH-CH_2-CH_2-CH=CH-CH_2\!\sim$

$+$

$\sim\!CH_2-CH=CH-CH_2-CH_2-CH=CH-CH_2\!\sim$

$\xrightarrow{\text{Oxygen}}$ $\sim\!CH_2-\underset{\underset{O}{|}}{CH}-\underset{\underset{O}{|}}{CH}-CH_2-CH_2-CH=CH-CH_2-CH_2-\underset{\underset{O}{|}}{CH}-\underset{\underset{O}{|}}{CH}-CH_2\!\sim$

$\sim\!CH_2-\overset{|}{CH}-\overset{|}{CH}-CH_2-CH_2-CH=CH-CH_2-CH_2-\overset{|}{CH}-\overset{|}{CH}-CH_2\!\sim$

The linear polymer formed by condensation between urea and formaldehyde can undergo further condensation to yield a cross-linked polymer:

$$H_2N - \underset{\underset{O}{\|}}{C} - NH - CH_2 - NH - \underset{\underset{O}{\|}}{C} - NH\sim\sim$$

condensation ↓

$$\sim\sim CH_2 - HN - \underset{\underset{O}{\|}}{C} - \underset{\underset{\underset{NH-\underset{\underset{O}{\|}}{C}}{|}}{\underset{CH_2}{|}}}{N} \qquad \underset{\underset{\underset{NH-\underset{\underset{O}{\|}}{C}-}{|}}{\underset{CH_2}{|}}}{\underset{\underset{O}{\|}}{\underset{|}{N}} - \underset{\underset{O}{\|}}{C} - NH\sim\sim}$$

(structure showing cross-linked urea-formaldehyde condensation product with NH–C(=O)–NH and CH$_2$ bridges, terminating in $-C(=O)-NH\sim\sim$)

1.2.2 Polymer Characteristics

If the atoms or groups within a polymer chain are highly ordered and the polymer undergoes some form of mechanical treatment such that the chains line up parallel to each other, crystallites will form. The bonds between the chains are weak, being secondary or van der Waals forces, but there are large numbers of such bonds and they are usually regularly spaced so that the polymer crystallises. Crystallisation results in a hardening effect upon the polymer and confers increased resistance to solvents and thermal action. Cross-linkage has a similar effect. In general, non-crystalline and/or non-cross-linked polymers are thermoplastic or mouldable.

The simple polymer derived from ethylene may be crystallised relatively easily. The presence of substituents on the double bonded carbons as in vinyl compounds, however, can prevent this crystallisation. In a typical vinyl compound, $CH_2=CHX$, a number of different substituent groups, X, may be attached to the carbon atom, see Table 1.1, and these groups can be arranged in three different ways on the polymer chain. If the side groups are randomly located about the carbon backbone (Figure 1.1a), the polymer is said to be atactic. When the side groups are all on the same side of the chain (Figure 1.1b), the polymer is described as isotactic. If, however, the side groups alternate from one side to the other in a regular manner (Figure 1.1c), the polymer is said to be syndiotactic.

Grouping, X	Compound	Polymer
Hydroxyl, -OH	Vinyl alcohol $CH_2 = \underset{\underset{H}{\vert}}{C} - OH$	$\\{CH_2 - \underset{\underset{OH}{\vert}}{CH}\\}_n$ Polyvinyl alcohol
Chlorine, -Cl	Vinyl chloride $CH_2 = \underset{\underset{H}{\vert}}{C} - Cl$	$\\{CH_2 - \underset{\underset{Cl}{\vert}}{CH}\\}_n$ Polyvinyl chloride, PVC
Acetate, CH_3COO-	Vinyl acetate $CH_2 = \underset{\underset{H}{\vert}}{C} - COOCH_3$	$\\{CH_2 - \underset{\underset{OOCCH_3}{\vert}}{CH}\\}_n$ Polyvinyl acetate, PVA
Phenyl, C_6H_5-	Styrene $CH_2 = \underset{\underset{H}{\vert}}{C} - C_6H_5$	$\\{CH_2 - \underset{\underset{C_6H_5}{\vert}}{CH}\\}_n$ Polystyrene
Methyl, CH_3-	Propylene $CH_2 = \underset{\underset{H}{\vert}}{C} - CH_3$	$\\{CH_2 - \underset{\underset{CH_3}{\vert}}{CH}\\}_n$ Polypropylene
Carboxyl, -COOH	Acrylic acid $CH_2 = \underset{\underset{H}{\vert}}{C} - COOH$	$\\{CH_2 - \underset{\underset{COOH}{\vert}}{CH}\\}_n$ Polyacrylic acid
Ester, $-COOCH_3$	Methyl acrylate $CH_2 = \underset{\underset{H}{\vert}}{C} - COOCH_3$	$\\{CH_2 - \underset{\underset{COOMe}{\vert}}{CH}\\}_n$ Polymethyl acrylate
Ether, $-OC_2H_5$	Ethyl vinyl ether $CH_2 = \underset{\underset{H}{\vert}}{C} - OC_2H_5$	$\\{CH_2 - \underset{\underset{OC_2H_5}{\vert}}{CH}\\}_n$ Polyethylvinyl ether

TABLE 1.1   Vinyl Compounds, $CH_2 = CHX$

Bulky side groups along the chain confer rigidity by inhibiting bending, but they will prevent, or at least severely restrict, crystallisation by interfering with the close approach of parallel chains necessary for formation of secondary bonds by van der Waals forces. Small side groups such as the hydroxyl groups in polyvinyl alcohol, for example, will not interfere in this manner and provided linear chains are formed, the polymer will crystallise easily. Non-crystalline structures, however, are generally found with atactic polymers containing large side groups, such as the chlorine atom or the phenyl group. Isotactic and syndiotactic polymers, even if bulky side groups are present, will usually crystallise. In general, the lower the regularity of the individual chains, the lower is the tendency towards crystallinity in the polymers.

The decreased solvent resistance of non-crystalline polymers, even those having rigid structures due to the presence of bulky side groups, appears to be due to the readier ingress of solvents and swelling agents between the polymer chains. As a result, non-crystalline polymers are more readily dissolved and are subject to swelling.

Vinylidene compounds (see Table 1.2) of the general formula $CH_2 = CXY$ are produced when two hydrogen atoms of the ethylene molecule are replaced by the substituents X and Y, where X and Y may be halogens, alkyl groups and more complex groupings. Vinylidene compounds can be polymerised to yield polyvinylidene resins and, as with the vinyl resins, the properties of the polymers vary with both the nature of the substituent groups and the geometry (i.e. positioning of the mers) of the polymeric backbone.

Copolymerisation can produce polymers having one of four different distributions of the individual mers along the chain (Figures 1.1 and 1.2a-d). Copolymerisation generally decreases the regularity of the polymer chains and, consequently, copolymers have a lower tendency to crystallise. This tendency is utilised in various copolymers to achieve optimum physical and chemical properties, e.g. the copolymerisation of vinyl chloride and vinyl acetate yields a polymer having a somwhat less rigid structure and improved solubility characteristics due to the incorporation of vinyl acetate. Similarly, vinyl chloride is often copolymerised with vinylidene chloride to reduce the crystallinity of polyvinylidene chloride and increase its flexibility.

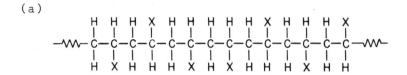

Figure 1.1   Arrangements of side groups in a simple
vinyl polymer $(CH_2-CHX)_n$

    (a) Atactic or random
    (b) Isotactic or positioned on the same side
    (c) Syndiotactic or alternating regularly

Grouping		Compound	Polymer
X	Y		
Chlorine	Chlorine	Vinylidene chloride $CH_2 = CCl_2$	$\left(CH_2 - \underset{\underset{Cl}{\|}}{\overset{\overset{Cl}{\|}}{C}}\right)_n$ Polyvinylidene chloride
Methyl	Methyl	Isobutylene $CH_2 = C(CH_3)_2$	$\left(CH_2 - \underset{\underset{CH_3}{\|}}{\overset{\overset{CH_3}{\|}}{C}}\right)_n$ Polyisobutylene
Methyl	Carboxyl	Methacrylic acid $CH_2 = \underset{\underset{}{}}{\overset{\overset{CH_3}{\|}}{C}} - COOH$	$\left(CH_2 - \underset{\underset{COOH}{\|}}{\overset{\overset{CH_3}{\|}}{C}}\right)_n$ Polymethacrylic acid
Methyl	Acetate	Methyl methacrylate $CH_2 = \underset{\underset{}{}}{\overset{\overset{CH_3}{\|}}{C}} - COOCH_3$	$\left(CH_2 - \underset{\underset{COOCH_3}{\|}}{\overset{\overset{CH_3}{\|}}{C}}\right)_n$ Polymethyl methacrylate

TABLE 1.2 Vinylidene Compounds, $CH_2 = CXY$

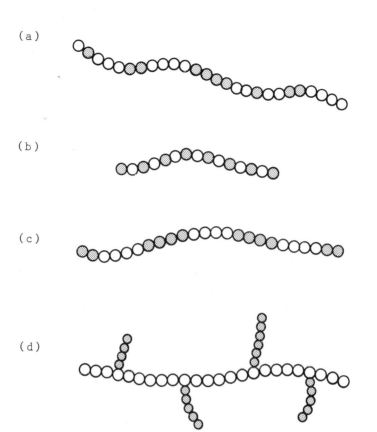

Figure 1.2 Distribution of monomers in copolymers

    (a) Random arrangement of separate mers along the chain
    (b) Regularly alternating mers along the chain
    (c) Block copolymer in which the blocks of individual mers alternate along the chain
    (d) Graft copolymer in which short chains of one mer attach at intervals along the chain of the other monomer

Cross-linking, as previously mentioned, provides bonds or ties between adjacent polymer chains, thereby restricting movement between the chains and modifying the mechanical behaviour. An obvious effect of the three-dimensional network resulting from cross-linking is the loss of thermoplasticity, and cross-linked polymers are referred to as thermosetting resins.

A thermoplastic or mouldable resin is one that exhibits a rigid or glass-like structure at low temperatures, but which becomes labile or plastic at elevated temperatures. The temperature at which this softening or loss of rigidity occurs is known as the glass transition temperature, T_g. This characteristic temperature is often well-defined, particularly in the absence of crystallinity, and permits the polymer to be moulded at elevated temperatures, i.e. temperatures above the T_g. Highly cross-linked polymers, however, do not exhibit a glass transition temperature. At elevated temperatures, these network polymers tend to degrade or depolymerise due to disruption of intrachain bonds by thermal energy.

Cross-linked polymers have, as a result of their rigid structures, good mechanical properties and high solvent resistance. It should be noted that variations also exist in the size of cross-linked networks, akin to the variable chain lengths found with addition polymers. Consequently, the structure of the polymer may consist of aggregates of highly cross-linked resin in a matrix of lower cross-link density, the two being held together by van der Waals forces. Cross-linked polymers consisting of long but bent chains with cross-links every one hundred or so atoms, and which are non-crystalline, are often elastomeric in character. Thus this type of polymer can be extended by over 100% (often up to 1000%) but will virtually instantaneously recover its dimensions on release of the strain. Finally, besides double bonds, other types of unsaturation or centres of reactivity such as strained rings are normally required for cross-linkage to occur. It is possible, however, to induce cross-linking in their absence. For instance, polyethylene, when irradiated, undergoes cross-linkage and loses its crystallinity to become slightly rubbery and more flexible. This phenomenon is utilised in radiation curing of paints which permits, amongst other advantages, the application of cross-linked polymers to heat-sensitive surfaces.

1.3 GENERAL PROPERTIES OF PIGMENTS

Pigments are finely dispersed particulate solids which impart certain desirable characteristics to paints. A prime requirement of pigments is their ability to add colour and opacity to paint films although other properties, such as their effects on the mechanical and durability characteristics of films, are also important.

Pigments can be either organic or inorganic compounds and their chemical composition must be such that they are able to withstand exposure to atmospheric conditions without substantial deterioration. Furthermore, they should be inert in paint media and be unaffected by the chemical nature of the substrate to which the paint is applied.

1.3.1 Particle Size and Related Properties

Most pigments and extenders used in paints and allied products are crystalline in nature. Tetragonal, rhombic and monoclinic crystal structures are typical. All crystals of a particular pigment produced under the same manufacturing conditions are similar in general shape and properties. A change in conditions can alter the crystal shape of a pigment and often modify its performance characteristics. Titanium dioxide, for example, is a pigment which, depending upon the method of manufacture, can have two different crystal forms, namely the anatase or the rutile forms, which exhibit markedly different performance properties (see Chapter 3). Non-crystalline pigments such as the carbon blacks, however, are also encountered in paint technology.

In the pigment manufacturing process individual pigment crystals, known as primary particles, generally form agglomerates composed of large numbers of individual pigment crystals. During paint manufacture, these agglomerates are mechanically reduced to smaller agglomerates or even primary particles, and become dispersed in the paint medium. The particle size of the dispersed pigment agglomerates or primary particles is of paramount importance in determining the performance of paint systems and, normally, a distribution of particle sizes is obtained during paint manufacture. This distribution is relatively uncontrollable but it is possible to ensure freedom from oversize particles (which can cause film 'bittiness') and a dispersion of particles in the paint medium that optimises pigment utilisation in the final coating.

Pigments	Specific gravity*	Oil absorption* (g/100g pigment)
Aluminium powder	2.55	-
Antimony oxide	5.75	13
Basic lead sulphate	6.4	12
Cadmium red	4.1	9
Cadmium yellow	4.2	44
Calcium plumbate	5.7	19
Carbon black (gas black)	1.95	144
Chrome green	2.84-5.01	13-23
Chromium oxide	4.76-5.2	10-18
Iron oxides, synthetic	4.90-5.15	22-39
Iron oxides, natural	3.4-5.1	14-40
Lead chrome (not orange)	5.84-6.40	12-20
Lead powder	11.40	-
Orange chrome	6.71-6.98	6.5-8.5
Phthalocyanine blue	1.64	40
Pigment Green B	1.47	54
Prussian blue	1.85-1.97	40-100
Red lead	8.95	5-12
Titanium dioxide, rutile	4.20	18-27
Titanium dioxide, anatase	3.88	
Toluidene red	1.40	35
Ultramarine blue	2.33	29
White lead	6.65	10
Zinc oxide	5.66	11-27
Extenders		
Barytes	4.25-4.5	10-15
China clay	2.6	30-60
Mica	2.8-2.85	75
Talc	2.65	30
Whiting	2.7	13-22

*Values for pigments from *Paint Trade Manual of Raw Materials and Plant*, (1973) p.67-68, Sawell Publications, London, (by kind permission of the Editor and Publishers).

TABLE 1.3   Physical Data on Pigments and Extenders

The optimum degree of pigment comminution and dispersion within a paint medium depends, to an extent, upon the end use of the coating. Thus, primers, undercoats and flat finishes can tolerate a lower degree of dispersion than paints with a high gloss requirement. Generally, gloss finishes are prepared with a maximum particle diameter of c. 5 μm since the gloss of such films would be reduced by the presence of larger particles. In contrast, undercoats and flat or eggshell finishes can tolerate larger particles, that is, a maximum of c. 20-40 μm in diameter. It is necessary, however, to restrict the proportion of larger diameter particles within a coating since they tend to introduce points of weakness into the film. Thus, films formed from vehicles in which there has been a poor comminution and dispersion of pigment particles would exhibit inferior durability characteristics when compared with those found when a more uniform dispersion was achieved.

Pigments and extenders vary considerably in the ease with which they can be dispersed in paint media, and the dispersion characteristics are determined by the size, shape and hardness of the particles. The texture of a pigment is most important in determining its practical applications since this characteristic influences the colour, tinting strength, oil absorption and opacity of the pigment, as well as the durability of the film in which it is dispersed.

Soft textured pigments disperse readily during paint manufacture and, generally, a lower particle size can be achieved with less expenditure of energy compared with hard textured pigments. This results in a higher surface area per unit volume of pigment with a concomitant greater oil absorption and opacity as well as a reduced tendency to hard, rapid settlement during storage of the liquid paint. Hard textured pigments are generally coarse and difficult to disperse whilst with high density materials, settlement problems can be encountered.

The specific gravities of pigments and extenders vary from approximately 1.4 for Toluidene red to 11.4 for metallic lead (Table 1.3) and this property has considerable importance in the context of the storage characteristics of the liquid paint. The rate of settlement of pigment particles through a liquid medium can be determined from Stokes' Law,

$$V = \frac{2gr^2 (D - d)}{9\eta} \qquad (1.1)$$

where V = velocity of settlement
       g = acceleration due to gravity
       r = radius of pigment particle
       D = specific gravity of pigments
       d = specific gravity of paint medium
       η = viscosity of paint medium

Although Stokes's law is only strictly applicable to spherical pigment particles, this equation enables an indication of the likely storage performance of a paint to be obtained at its development stage.

Pigments with a low specific gravity generally have low settling tendencies in paints, high oil absorptions and high tinting strengths. Pigments of high specific gravity are more prone to rapid settlement and, on a weight basis, contribute lower opacities, that is, lower hiding powers, to paint films.

The wetting of pigments and extenders by paint media is a complex phenomenon which has an important bearing on the rheological (flow) characteristics of paints. A paint medium or binder is generally considered to wet a pigment mass by filling the voids between the particles, filling fissures on individual particles and by absorption over the pigment surface. It is this absorbed layer of paint medium around the pigment that is important in determining the rheological characteristics of pigment-media suspensions.

A rough measure of the wetting characteristics of pigments can be obtained from a knowledge of their oil absorptions (Table 1.3). The oil absorption of a pigment or extender is defined as the minimum quantity of medium (the determination traditionally being performed using linseed oil) required to convert a given mass of dry pigment to a coherent uniform paste. This characteristic of pigments depends chiefly on the particle size, that is, the surface area of the pigment, with lower particle sizes being associated with increased oil obsorption, although the particle shape is also a factor in oil absorption.

Another important attribute of pigments is their tinting strength, this being the amount of a coloured pigment required to tint (that is, colour or stain) a given weight of a white pigment to produce a given shade. Tinting strengths are always relative to a standard sample of the pigment under test and, for two samples of the same pigment, the tint strength is a measure of the differences in particle size and distribution. Generally, pigments of low particle size have a high tint strength and, for this reason, organic pigments normally have higher tinting strengths than inorganic pigments of similar shade.

For white pigments, the quantity to produce a given colour when mixed with a given weight of a coloured pigment is a measure of its reducing power. Smaller amounts of a high opacity white pigment are required to obtain a given colour compared with low opacity white pigments, this being attributable to the greater scattering power of the higher opacity pigments, see Section 1.3.3.

1.3.2 Colour

The ability of pigments to impart colour is of fundamental importance and is attributable to the selective absorption of certain wavelengths of light by the pigment. The term 'colour' refers to all of the sensations aroused in the brain of an observer by the response of the retina of the eye, and its attached nerve mechanisms, to the radiant energy of certain wavelengths and intensities of light. For example, an article has a particular colour, such as red, due to the fact that of all the incident light striking the surface of the object, a certain proportion is absorbed and the rest is reflected back and detected by the retina of the eye. This reflected light consists primarily of those wavelengths which give rise to the sensations of red (Table 1.4). Objects which totally absorb all incident light appear black, and, conversely, objects which reflect all light appear white.

Colour absorbed	*Wavelength (nm)*	*Approximate colour of absorbant*
Violet	380-436	Yellow-Green
Blue	436-495	Yellow
Green	495-566	Red-Purple
Yellow	566-589	Violet
Orange	589-627	Green-Blue
Red	627-780	Blue

TABLE 1.4   Relation between Light Absorbed and Colour

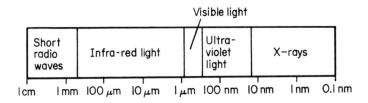

Figure 1.3 Radiant energy spectrum of the sun

Visible light is that portion of the sun's radiant energy spectrum (Figure 1.3) covering the region of 380 nm to 720 nm. Newton demonstrated by a classic series of experiments in 1670 that this visible white light could be split by a prism into a spectrum of colours. These colours are red, orange, yellow, green, blue, indigo and violet, together with numerous intermediate shades. The colours which appear in the spectrum are known as spectral colours and no purer colours than these are physically possible. In addition, there are a series of purple colours which do not appear in the spectrum of white light but which consist of mixtures of the extreme red and the extreme blue end of the spectrum. Such colours are known as non-spectral.

Colours can be described scientifically in terms of three characteristics, namely, hue, tone and intensity. The characteristic 'hue' describes that attribute of a colour which determines what kind of colour it is, for example, whether red, green or blue etc. Thus, hue is the dominant wavelength of the light reflected from a surface. All colours can be ranked on an arbitrary light-to-dark scale by an observer and the 'tone' of a colour is its position on such a scale. 'Intensity' is the characteristic by which the brilliance of a colour is seen and an intense colour is free from adulteration with other colours or dilution by white light, and is called a pure colour.

As previously stated, white light can be split into seven principal spectral colours, these being red, orange, yellow, green, blue, indigo and violet and clearly it is possible to recombine light having these wavelengths to produce white light. However, white light can also be obtained by blending only three colours, one from each end of the spectrum namely blue and red, and one from the middle of the spectrum, green. These three colours if projected together on a screen will produce white light and they are designated additive primary colours. Nearly every colour can be matched by combining together suitable proportions of these additive primaries and this method of colour formation is known as the additive system.

This system forms the basis of certain types of colour measuring instruments and with such instruments the colour of a given sample is determined by measuring the amounts of the three primary colours required to match it. The subject of colour measurement is discussed in greater depth in Chapter 9.

An alternative approach to obtaining colour, and the method by which most colours are obtained in practice, is the subtractive system. When white light is incident upon a coloured object, such as a surface carrying a pigmented coating, some wavelengths are preferentially absorbed whilst others are reflected. For example, a blue pigment will absorb the red and yellow components in white light and a yellow pigment will absorb the blue, indigo and violet components of white light. Green is the only colour that neither pigment absorbs and therefore when the two pigments are mixed, a green colour is obtained. Thus, colours in the subtractive system are derived from the removal or subtraction of certain wavelengths from the continuous wavelength spectrum of white light.

In subtractive mixing of colours, three transparent overlapping areas of the subtractive primary colours, namely red, yellow and blue, overlaid on a white background will result in total absorption of the incident white light from the background and produce black. This method of colour formation provides another means of colour measurement (see Chapter 9).

It should be noted, however, that no colour filter or pigment is pure. A blue pigment, for example, will absorb red and yellow but it will also absorb a proportion of the blue light incident upon it. Consequently, a wide variety of shades of green may be achieved by mixing blue and yellow pigments which respectively absorb greater or lesser amounts of the green component of the incident white light. Furthermore, the colour absorbing characteristics of pigments, and therefore their apparent colours, may alter with time due to chemical degradation, whilst colour absorption is often modified by the pigment dispersion medium.

1.3.2.1 *Colour fastness of pigments*

The ability of pigments to maintain their colour during exposure to sunlight is of considerable importance and this attribute of pigments is known as light fastness. Typically, pigments are found to discharge their colour, or fade, during service exposure although changes to darker shades are also encountered.

The light fastness of a pigment within a paint medium is dependent both upon its chemical composition and on such factors as its concentration within the coating and the nature of the paint medium. Of particular importance is the pigment concentration, since the use of a coloured pigment at low concentration in admixture with other pigments generally reduces its light fastness compared with the value at full strength, that is, in its unadulterated form. Because of these modifying factors it is impossible to categorise pigments into precise light fastness classifications although an attempt has been made in Chapter 3 to give an indication of the likely performance of individual pigments.

It was stated earlier that the colour of a pigment is determined by its selective absorption of white light. This ability to absorb certain wavelengths of light can be attributed to the presence, within the molecular structure of pigments, of certain specific electron configurations associated with ions, molecules and complexes. In particular, the electrons of the outer electron shell of atoms within molecules are most free to change their energy levels and it is the ability of these outer electrons to absorb and re-emit radiant energy or light that is responsible for the colour of pigments.

With inorganic, coloured pigments, the presence of imperfections (disorder) within the lattice of the pigment crystal structure is of importance in determining colours. For example, synthetic iron oxide reds when calcined at temperatures below $700^{\circ}C$, have a higher degree of disorder and a lighter colour than when calcined above this temperature. Since the bond energies in crystal lattices are high, considerable external energy is required to modify their structure and, consequently, the relatively low energies of light generally have little effect on the colour stability of inorganic pigments.

The colour of organic pigments is attributable to the presence within the molecule of certain chemical groups known as chromophores and examples of such groupings are the azo, -N=N-, nitroso, -N=O, and carbonyl, $>$C=O, groups. Other groups, known as auxochromes, are also a necessary requirement in the molecular structure of organic pigments and these serve to intensify the colouring effects of the chromophore groups. Some typical auxochromes are the amino, -NH$_2$, and hydroxyl, -OH, groups. The electrons in the chromophore and auxochrome groups within organic pigments are less rigidly bonded than those in inorganic crystals. Therefore, organic pigments can respond to lower energy radiation such as light more easily than the inorganic pigments and tend to fade or discolour. This is, however, something of a simplification for there are examples of organic pigments with a very high degree of light fastness, particularly those containing complex chromophore groupings.

Colour changes can also be induced in pigments by chemical attack by the environment to which they are exposed, e.g. blackening of lead pigments in sulphur-rich atmospheres and the discolouration of Prussian blue on alkaline substrates. The chemical composition of the pigment is therefore an important factor in determining its chemical resistance and colour or light fastness. Details concerning individual pigments are given in Chapter 3.

These chemically induced colour changes are found to occur concomitantly with light induced changes during exposure. Any marked tendencies to such defects, however, can generally be avoided during the formulation stage of a pigmented coating by suitable selection of pigmentary materials.

1.3.3 Opacity

The ability of pigments to confer opacity to paint films is of particular importance since it is this characteristic which determines the ability of the film to obliterate or hide the substrate to which it is applied.

When light falls on a pigmented paint film, part is reflected back whilst some enters the film. The light which is reflected back produces the sensations of gloss, with high reflectance being characteristic of glossy films and low reflectances characteristic of flat (or matt) films. The light which enters the film is subject to scattering and absorption by the pigment particles within the film, and it is the degree to which these effects occur that influences the opacity of the film. Light that does not interact with the pigment can be reflected back from the substrate to the film-air interface (undergoing further successive interactions with the pigment on the way back through the film). If this effect occurs, then the substrate will be visible through the paint film.

Opacity, or hiding power, therefore, is dependent upon both the phenomenon of scattering and the absorption of light by pigment particles. With black and strongly coloured pigments, opacity is mainly due to the absorption of light whilst with white pigments absorption is minimal and scattering is the dominant effect. Other factors such as pigment loading, particle size and film thickness also influence the hiding power of paint films.

Scattering power is dependent upon the refractive indices of the pigment and of the dried binder film in which it is dispersed (Table 1.5). The scattering power of white pigments can be determined from the expression:

$$\text{Scattering power} = \frac{(n-1)^2}{(n+1)^2} \quad (1.2)$$

where n is the ratio of the refractive indices of the pigment, n_p, to that of the binder in which it is dispersed, n_b.

This expression is known as the Fresnel reflectance function and some values for white pigments in an alkyd binder are given in Table 1.6. These values clearly demonstrate the increasing amount of light scattering that occurs as the refractive index of white pigments increases. The last group of materials given in Table 1.6 are examples of extenders and, due to their low refractive indices, they exhibit poor opacity in oleoresinous and other solvent soluble media. However, particularly in emulsion paints of high pigment volume concentration, extenders can act as opacifying pigments, since the particles in the dried film are predominantly surrounded by air, which has a refractive index of 1.0003, giving a relatively high value of n in Equation 1.2.

White pigments (most commonly titanium dioxide) are generally included in paint formulations to confer opacity, since they have a higher refractive index than most coloured pigments, with consequently greater scattering power. It must be remembered, however, that the ability of coloured pigments to confer hiding power is not solely

Pigments and extenders	Refractive index of pigments and extenders, n_p
Titanium dioxide, rutile	2.76
Titanium dioxide, anatase	2.52
Antimony oxide	2.09
Zinc oxide	2.08
Barytes	1.64
Whiting	1.58
Talc	1.40

Resins	Refractive index of resins, n_b
Linseed oil	1.48
Alkyd resins	1.50-1.60
Phenolic resin	c. 1.65
Vinyl resins	c. 1.6
Nitrocellulose resins	c. 1.5

TABLE 1.5   Refractive Index of Pigments, Extenders and Resins in Air

Pigment	Scattering power
Titanium dioxide, rutile	0.0788
Titanium dioxide, anatase	0.0568
Antimony oxide	0.0220
Zinc oxide	0.0213
Extenders	
Barytes	0.0008
Whiting	0.0001

TABLE 1.6   Light Scattering of White Pigments and Extenders in an Alkyd Resin (n_b = 1.55)

dependent upon scattering of light but is also influenced by their absorption characteristics. As previously stated, the hiding power of a paint is influenced by the nature of the interface between the pigment and its binder matrix and, in particular, on its extent. The hiding power of pigments also depends upon the particle shape and hence the method of manufacture. Film opacity can be increased by reducing the size of the pigment particles and so increasing their reflective surface area, although for white pigments this effect may be reversed if the size falls much below half the wavelength of visible light, namely 0.2-0.35 μm. However, black pigments and highly saturated coloured pigments retain good hiding power even at very small particle sizes, since in these instances absorption rather than scattering is the dominant factor.

During paint manufacture, opacifying pigments are dispersed in the binder to a particle size which imparts the required degree of opacity to the film. However, extender materials are frequently added to paints and these, by their modifying effect on the total particle size distribution of the pigment, can influence the opacity of the film. For example, extenders of fine particle size increase 'hiding' by separating the pigment particles, whilst extenders of large particle size decrease hiding by crowding the pigment particles together.

The opacity of a film will also increase with its pigment content since the number of pigment surfaces at which scattering can occur is increased and, in a similar fashion, opacity also increases with paint film thickness.

In paint technology the measurement of the covering power of paints, that is, the dry thickness of a paint needed to obliterate the substrate, is of considerable practical importance. An indication of a paint's covering power can be obtained by determining its contrast ratio after application at a predetermined wet thickness. This figure is obtained from the ratio of the reflectance of white light from the film when applied over a black substrate, R_b, and its reflectance when applied over a white, substrate, R_w:

$$\text{Contrast ratio} = \frac{R_b}{R_w} \quad (1.3)$$

It is normally found that a paint film with a contrast ratio of 0.98 is required to completely obscure a coloured substrate. However, for certain applications, for example, low cost paint systems, and with particular organic pigment combinations, lower contrast ratios are frequently accepted. Low opacity is frequently found with films pigmented with organic pigments because their addition rates are necessarily low due to their effects on the rheological properties of the paint and the often rather high cost of such pigments.

Another effect frequently observed when applying paints is a change of opacity, normally to lower values, when the paint films dry. This effect is normally experienced with solvent-containing paints (see Section 2.1) and is attributable to the low refractive indices of solvents within the coating producing anomalously high values of n in Equation 1.2. As the coating dries and the solvent is lost from the film by evaporation, the value of n approaches its theoretical value for the particular pigment-resin combination.

1.4 BIBLIOGRAPHY

Technology of Paints, Varnishes and Lacquers, C.R. Martens, Reinhold, New York (1968).

Organic Polymer Chemistry, K.J.Saunders, Chapman and Hall, London (1973).

Pigments: An Introduction to their Physical Chemistry, D. Patterson, Elsevier, Lausanne (1967).

Pigment Handbook, Vols. I, II & III, T.C. Patton, Wiley, New York (1973).

2 Polymers used in paints

A wide variety of binders or, as they are alternatively known, polymers, resins or vehicles, are used in paints. Film formation occurs when the liquid binder converts to a solid. This liquid-solid conversion may involve a chemical reaction such as an addition or a condensation polymerisation process, an oxidative polymerisation reaction or the evaporation of a solvent in which the polymer is carried. These various reactions are collectively referred to as paint curing processes and, clearly, the nature of the polymeric component or the binder of the paint will determine the actual reaction involved in the curing or drying of a paint. In this chapter, the common binders used in paint technology are discussed.

2.1 CONVERTIBLE AND NON-CONVERTIBLE COATINGS

Paints can be subdivided into two broad categories, convertible and non-convertible coatings. A convertible coating is a paint in which the binder is either a polymer precursor, a monomer or a partially polymerised material. Upon addition of a suitable initiator, or exposure to radiant energy such as ultraviolet, infrared or nuclear radiation, the monomeric or partially-polymerised component of the paint undergoes reaction and polymerisation occurs. This polymerisation may be an addition or a condensation reaction or, often, a cross-linking reaction such that the binder is converted from a liquid or soluble state into an insoluble solid material. This solid resin constitutes the binder of the paint film. Thus the curing or drying of a convertible coating involves some form of chemical reaction which converts the liquid paint into a solid film.

Non-convertible coatings, in contrast, do not undergo curing or chemical conversion reactions when they dry. Film formation or drying of the paint involves loss or evaporation of a volatile solvent or a dispersion medium and concomitant deposition of a solid material. If the polymeric component or binder is soluble in a volatile solvent, then simultaneous solvent evaporation and formation of a uniform, continuous film requires control of the evaporation rate through selection of the solvent blend. This ensures that the deposit is both uniform and continuous whilst containing few macrovoids.

2.2 OLEORESINOUS BINDERS

Amongst the oldest and possibly still the most successful binders are the vegetable oils, notably linseed oil. Linseed oil which is obtained from the flax plant is a mixture of long chain fatty acids, mainly linolenic acid, linoleic acid and oleic acid. Other oils that are also used include tung, olive, soya, castor, coconut, cottonseed and oiticica.

These oils may be broadly classified as drying, non-drying and semi-drying. Drying oils, such as linseed, tung and oiticica, contain unsaturated fatty acids and, on exposure to air, they absorb oxygen and harden. This hardening and film-forming process involves a complex reaction, known as oxidative polymerisation, in which oxygen adds to the carbon-carbon double bonds and then adjacent chains of the acids are linked by oxygen bonds. Non-drying oils such as olive, castor, coconut and cottonseed, are composed of saturated fatty acids and cannot undergo this type of reaction. However, castor oil can be dehydrated to yield an unsaturated drying oil, 'dehydrated castor oil' (D.C.O.), which has a wide usage. Semi-drying oils, for example, soya bean oil, contain fewer unsaturation sites than the drying oils. Consequently, these oils can undergo a degree of oxidative polymerisation but the reaction is slower and less complete that for the drying oils. Non-drying and semi-drying oils therefore are not used as binders *per se*, rather finding use in the manufacture of more complex binders, notably the alkyd resins (see Section 2.3).

Drying oils form films only slowly and the film tends to be somewhat soft, so the addition of driers, for example, salts of lead, cobalt or manganese, is required to accelerate the curing reaction.

The performance of paints derived from oil-based vehicles can be markedly improved by reacting the oil with natural or synthetic resins to yield oleoresinous varnishes. These varnishes are classified as short-oil, medium oil or long oil depending on the oil to resin ratio. Short oil vehicles contain approximately 75 litre per 50 kg resin, medium oil contains 75-150 litre per 50 kg resin, whilst long oil vehicles contain more than 150 litre per 50 kg resin. The resins used in varnish manufacture include rosin, coumarone, indene and kauri resins. In general, short oil vehicles yield rapid drying, hard, brittle films whilst long oil vehicles yield softer, more flexible films with the medium oil vehicles having intermediate properties. Varnishes also cure by oxidative polymerisation, or auto-oxidation, but the paint drying time is markedly shorter than that of the unmodified drying oils.

2.3 ALKYD RESINS

Alkyd resins are produced by condensation polymerisation of esters formed from the reaction between polycarboxylic acids and polyhydric alcohols. The simplest esterification reaction is that between a fatty acid, such as acetic acid, and an alcohol, for example ethyl alcohol or, more correctly, ethanol:

$$CH_3.COOH + C_2H_5.OH \longrightarrow CH_3.COOC_2H_5 + H_2O$$

Acetic acid   Ethanol   Ethyl acetate

More complex esters are formed between polyhydric alcohols, that is, alcohols containing more than one hydroxyl group in the molecule, and polycarboxylic acids, that is, acids containing more than one carboxylic acid grouping per molecule. Glycerol and phthalic acid or phthalic anhydride are often used to prepare alkyd resins. The first stage in the process is esterification of the alcohol:

In the derived ester, each phthalic acid molecule esterifies two different glycerol molecules. The commercial alkyd manufacturing process is based upon this esterification process but differs in that esterification and condensation polymerisation of the constituents is allowed to proceed. Furthermore, the reaction mixture contains a modifying fatty acid or an oil in addition to the polyhydric alcohol and polycarboxylic acid. Fatty acid modified alkyds are prepared by heating the constituents together in a special reaction vessel or kettle at about $240°C$ until esterification is complete, this procedure being known as the fatty acid process. The fatty acids used for modifying the alkyds are derived from the oils commonly used in oleoresinous vehicles, for example, linolenic or linoleic acid from linseed oil.

An alternative procedure is used to manufacture the oil modified alkyds, namely the alcoholysis process, since the oils are generally insoluble in the initial polyhydric alcohol-polycarboxylic acid mixture. In the alcoholysis process, the oil and polyhydric alcohol are mixed and heated at $240-260°C$ in the presence of a suitable catalyst such as calcium hydroxide. The alcoholised reaction product, a monoglyceride, is then reacted as in the fatty acid process with the dibasic acid to yield the oil-modified alkyd resin. The most widely used oils for the preparation of semi-drying and drying oil-modified alkyds are soya bean and linseed oils respectively, although other oils, are also used. The commoner polycarboxylic acids and polyhydric alcohols (or polyols) used in alkyd resin manufacture are given in Table 2.1.

Polycarboxylic acids	*Polyhydric alcohols*	*Oils*
Phthalic anhydride	Glycerol	Soya bean
Isophthalic acid	Pentaerythritol	Linseed
Orthophthalic acid	Sorbitol	Castor
Maleic acid	Trimethylolethane	Coconut
Fumaric acid	Trimethylolpropane	Tung
Adipic acid	Ethylene glycol	
Sebacic acid	Propylene glycol	

TABLE 2.1   Raw Materials used in Alkyd Resin Manufacture

Alkyd resins fall into three broad categories, namely drying, semi-drying and non-drying alkyds, depending on the oil used in conjunction with the alkyd resin. These categories may be subdivided further on the basis of oil length, this being the amount of oil, or fatty acid, in the alkyd resin.

Short drying oil alkyds (up to 45% oil length) are soluble only in aromatic solvents and are cured by high temperature, that is, baking or stoving. Medium oil alkyds (45-60% oil length) are soluble in aromatic or aromatic-aliphatic solvent mixtures and these can be cured by either air-drying or high temperature processes.

Non- and semi-drying oil alkyd resins of these oil lengths are primarily used in conjunction with other film forming resins, such as amino, melamine or nitrocellulose resins. Medium oil length alkyds are frequently used as binders in certain types of quick air-drying finishing systems.

The long oil alkyds (60-80% oil length) are normally prepared from drying oils (or their fatty acids), and resins of these oil lengths are completely soluble in aliphatic solvents. They are used primarily as binders for decorative finishing systems, and film formation is normally by air drying.

2.3.1 Film Formation

The conversion mechanism of a liquid alkyd resin to a solid film is dependent upon the fatty acid residues present in the resin. Non-drying oil alkyds do not readily form films and, as such, they are mainly used as plasticisers for other binders. Semi-drying and drying oil alkyds, however, can form films through oxidative polymerisation in a similar manner to that of the simple oils and oleoresinous binders. Alkyd resins have greater molecular weights than the oil precursors. Consequently, fewer cross-linkages, and a correspondingly shorter curing time, are required for film formation.

As with the oleoresinous binders, curing of alkyd resins to form films involves two steps. There is an initial oxidation (by atmospheric oxygen) of double bonds within the fatty acid residues in the alkyd resin. This primary oxidation is followed by decomposition of the oxidation products which initiates cross-linkage. The detailed mechanism of the initial oxidation and the subsequent cross-linkage reactions is not fully understood but would appear to differ for conjugated and non-conjugated fatty acids. Fatty acids containing conjugated double bonds possess two double bonds separated by a single carbon-

carbon bond, -C=C-C=C-, whilst in non-conjugated acids, the double bonds are more widely spaced, e.g., -C=C-C-C=C-. The presence of conjugation, as in the fatty acids of tung oil, confers greater reactivity and faster film formation than the absence or a reduced degree of conjugation, as in the fatty acids of linseed oil. Aerial oxidation of non-conjugated fatty acids yields hydroperoxides whilst the conjugated acids appear to form peroxides.

Cross-linkage, and therefore film formation, tends to be slow even in highly conjugated fatty acids. Consequently, driers are added to the system (see Section 4.3). These driers catalyse the decomposition of the hydroperoxides or peroxides so that cross-linkage is accelerated. Oxidation and particularly cross-linkage may continue after the initial cure so that excessive cross-linking occurs. This often leads to deterioration of the film and in alkyd resins, where conjugated hydroperoxides can form, film yellowing may be observed.

2.3.2 Modified Alkyd Resins

Alkyd resins are polyesters and, as such, they are susceptible to degradation by alkalies. The chemical resistance and other characteristics can be improved, however, through modification of the alkyd by incorporation of other materials. Typically, chlorinated rubber improves corrosion and fire resistance and accelerates the drying rate, cellulose nitrate improves hardness and drying rate, amino resins increase the hardness and alkali resistance, and silicone resins improve water and heat resistance.

The above additions to the alkyd binder are made by blending or physically combining the second resin with the alkyd. Chemical additions of other components are made to alkyds by copolymerising the alkyd with vinyl type monomers such as styrene, vinyl toluene, or with isocyanate groups to form urethane alkyds. This copolymerisation of alkyd resins may be performed during manufacture of the alkyd resin or the prepared alkyd resin may be used as the reactant in a separate process.

The most widely applied process is styrenation of alkyds in which the alkyd resin is reacted with styrene in the presence of a peroxide catalyst. Reaction and addition of styrene occurs at double bonds in the fatty acid residues of the alkyd resin. Styrenated alkyds have improved drying rates and chemical resistance compared with conventional alkyds.

2.4 AMINO RESINS

Amino resins are formed by condensation reactions between aldehydes ($R \cdot CHO$) and amines ($R' \cdot NH_2$) or amides ($R' \cdot NH \cdot R''$). The amino resins of greatest commercial importance are the urea-formaldehyde and the melamine-formaldehyde polymers.

Preparation of the urea-formaldehyde resin involves several stages. There is an initial reaction under alkaline conditions which yields a mixture of methylolureas:

$$\underset{\text{Urea}}{\begin{array}{c} NH_2 \\ | \\ CO \\ | \\ NH_2 \end{array}} + \underset{\text{Formaldehyde}}{\begin{array}{c} H-C-H \\ \| \\ O \end{array}} \longrightarrow \underset{\text{Monomethylolurea}}{\begin{array}{c} NH \cdot CH_2OH \\ | \\ CO \\ | \\ NH_2 \end{array}} + \underset{\text{Dimethylolurea}}{\begin{array}{c} NH \cdot CH_2OH \\ | \\ CO \\ | \\ NH \cdot CH_2OH \end{array}}$$

Under acidic conditions, the methylolureas undergo condensation with urea to form methylene compounds of the type:

$$\begin{array}{c} NH \longrightarrow CH_2 \longrightarrow NH \\ | \qquad\qquad\qquad\qquad | \\ CO \qquad\qquad\qquad\quad CO \\ | \qquad\qquad\qquad\qquad | \\ NH \cdot CH_2OH \qquad\quad NH \cdot CH_2OH \end{array}$$

but these compounds can undergo further reaction to yield linear polymers of the general formula:
$HOCH_2 \dashv NH \cdot CO \cdot NH \cdot CH_2 \dashv_n NH \cdot CO \cdot NH \cdot CH_2OH$.

Similar condensation reactions occur between melamine (triaminotriazine) and formaldehyde to form methylolmelamines.

Melamine

These methylolmelamines can undergo further condensation to form linear polymers.

The unmodified urea-formaldehyde and melamine-formaldehyde resins are insoluble in common solvents so they are usually modified with butanol, which is added to the initial mixture. This yields a linear polymer in which some of the methylol groups are butylated:

$$-NH \cdot CO \cdot NH \cdot CH_2OH \xrightarrow{C_4H_9OH} -NH \cdot CO \cdot NH \cdot CH_2OC_4H_9.$$

These butylated urea-formaldehyde (amino) resins are soluble in the common solvents and are commonly used in coating formulations in conjunction with short and medium oil length alkyd resins. Alkyd-butylated amino resin blends give coatings with good flexibility and adhesion as well as satisfactory gloss, reasonable chemical resistance and a rapid curing rate.

It should be noted that both non-butylated urea-formaldehyde and melamine formaldehyde resins will undergo cross-linking reactions under acidic conditions. These high molecular weight cross-linked or network polymers also have great industrial importance for moulding applications.

2.5 VINYL RESINS

There are a large number of vinyl compounds having the general formula of $CH_2=CHX$ where X may be Cl, CH_3COO, OH and many others. Only two polymerised vinyl compounds, namely the chloride and the acetate, and their copolymers, have achieved major importance in the coatings field. Poly (vinyl alcohol), poly (vinyl formal) and poly (vinyl butyral) also have importance in other fields, as have poly (vinyl chloride) and poly (vinyl acetate), as well as in surface coatings.

2.5.1 Poly (Vinyl Chloride), PVC

Poly (vinyl chloride) is a colourless rigid material that has a relatively high T_g (80°C) and is insoluble in the common solvents. Furthermore, it is virtually unaffected by acids, alkalies and oxidising agents, although it is sensitive to heat and light.

The polymer is manufactured by three principal preparative routes, namely, bulk, suspension and emulsion polymerisation. Bulk polymerisation involves heating vinyl chloride with an initiator such as benzoyl peroxide at 60°C for 10-12 hours in a sealed rotating drum containing stainless steel balls. The particle size of the poly (vinyl chloride) is determined by the drum rotation speed and the ball diameter. Suspension polymerisation involves polymerisation of vinyl chloride suspended in water at 50°C under pressure in the presence of a suitable initiator. The reaction yields a slurry of the polymer in water from which it is filtered and dried. The shape of the polymer particles, e.g. spherical, irregular, etc., is determined by the suspension agent used in the process. Emulsion polymerisation is also performed in a pressure vessel at about 50°C but the vinyl chloride is present as an emulsion with water. The reaction is faster (1-2 hours) but produces a latex which has to be spray-dried to yield very small spherical particles.

A fourth processing route is also used for poly (vinyl chloride), namely solution polymerisation. In this process, the monomer is dissolved in a solvent and, after catalyst addition, polymerisation proceeds in the solvent medium. No isolation of the resultant PVC is necessary, since the solvent used for polymerisation is the same as that used for the final resin.

It should be noted that the polymerisation processes outlined above are used for all the vinyl compounds.

Poly (vinyl chloride) or PVC is a rigid, colourless, polymer possessing excellent chemical resistance to acids, alkalies, aqueous media and oxidising agents. The pure PVC resins are linear thermoplastic polymers having the structure:

$$\sim\!\!CH_2\!-\!\underset{\underset{Cl}{|}}{CH}\!-\!CH_2\!-\!\underset{\underset{Cl}{|}}{CH}\!-\!CH_2\!-\!\underset{\underset{Cl}{|}}{CH}\!-\!CH_2\!-\!\underset{\underset{Cl}{|}}{CH}\!\sim$$

that is, they have an isotactic structure (see Section 1.2.2), and therefore have a tendency to crystallise. This would account for the limited solubility of PVC in the common solvents. It will dissolve, however, in oxygen-containing solvents, such as ketones, ethers and aromatic nitro-compounds, as well as chlorinated solvents, when interaction between solute and solvent appears to occur.

In general, the poor solubility characteristics of pure PVC resin limits its applications in surface coatings although it is used in dispersion systems. If the resin is dispersed in a solvating medium such as a hydrocarbon, the dispersion is known as an organosol. Dispersions in plasticisers are referred to as plastisols. When the dispersions are sprayed onto the substrate, discrete particles of the resin are deposited and, when the substrate is baked, these particles coalesce to form a continuous coating.

Vinyl chloride is used more commonly, however, as a copolymer with vinyl acetate.

2.5.2 Poly (Vinyl Chloride-Vinyl Acetate)

Copolymerisation of vinyl acetate with vinyl chloride yields polymers of decreased rigidity and greater solvent solubility but which retain a high degree of chemical resistance. The properties of the copolymer are determined by the vinyl chloride to vinyl acetate ratio and, commonly, ratios of 80:20 to 90:10 are used for surface coating binders.

The copolymers are prepared by techniques essentially the same as for the vinyl chloride homopolymer, although suspension and solution polymerisation processes are used most widely. Solution polymerisation is used to a greater extent since it offers certain technical advantages, notably more efficient input of the monomers to ensure uniform copolymer composition. Polar groups are often incorporated into the copolymer to promote adhesion to metallic substrates by low additions ($c.$ 1%) of carboxylic acids, such as maleic acid, to yield terpolymers. A similar effect is achieved by copolymerising pure vinyl chloride with maleic acid.

Poly (vinyl chloride-vinyl acetate) is a linear thermoplastic copolymer having a somewhat lower tensile strength than PVC. Coatings based on the copolymer are, like the homopolymer PVC, non-converting and they are are applied by a spray procedure. Film formation is generally by air drying, i.e. deposition from solvent solution, although baking techniques are also sometimes used.

The copolymer has the structural formula:

$$\left[\begin{array}{c} Cl \\ | \\ CH-CH_2 \end{array} \right]_n \left[\begin{array}{c} CH_3 \\ | \\ C=O \\ | \\ O \\ | \\ CH-CH_2 \end{array} \right]_n$$

n = 80-90% n = 20-10%

and, in general, lower molecular weight resins are used as paint binders. The copolymer is more soluble in a wider range of solvents than PVC.

2.5.3 Poly (Vinyl Butyral)

Poly (vinyl butyral), which is used in metal pretreatment primers (see Chapter 7), is prepared from poly (vinyl alcohol). Hydrolysed poly (vinyl alcohol) is suspended in alcohol, to which butyraldehyde and a catalyst (sulphuric acid) is added, the mixture being heated to 80°C for several hours. After reaction, addition of water precipitates the copolymer which has the structural formula:

$$\left[\begin{array}{c} CH-CH_2-CH \\ | \quad\quad\quad\quad | \\ O-\;\;CH\;\;-O \\ | \\ C_3H_7 \end{array} \right]_n$$

The copolymer forms tough, flexible films which exhibit high adhesion to metallic substrates, even under prolonged immersion. The films are attacked by acids and strong alkalies but they are resistant to most oils and aliphatic hydrocarbons.

2.5.4 Poly (Vinylidene Chloride)

Vinylidene chloride, $CH_2=CCl_2$, polymerises to a very regular, dense crystalline resin with a high melting point (220°C). As a result, poly (vinylidene chloride) exhibits a very low permeability to gases, water vapour and aqueous media although its films have poor flexibility. The resin is usually manufactured by a suspension polymerisation process. The crystalline nature of poly (vinylidene chloride) limits its applications and it is commonly used as a copolymer with vinyl chloride which yields a resin having greater film flexibility and a somewhat lower softening point (140°C). Monomer ratios of 85:15, vinylidene chloride: vinyl chloride, are the most widely used and find application where high resistance to aqueous media is required.

2.6 ACRYLIC RESINS

Acrylic resins, the most widely used being poly (methyl methacrylate), are polyvinylidene compounds having the general formula (see Table 1.2):

$$-[CH_2-CXY]_n-$$

(where X commonly may be H, CH_3, C_2H_5 and Y is usually COOH, $COOCH_3$)

and include polymers derived from acrylonitrile ($CH_2\!=\!CH\text{-}CN$) and acrylamide ($CH_2\!=\!CH\text{-}CONH_2$).

The resins used in surface coatings are derived from the esters of acrylic and methacrylic acid and of these, poly (methyl methacrylate), PMMA, also has great industrial importance as a moulding resin, e.g. Perspex, Lucite, Plexiglas. The resin is prepared by a vinyl-type polymerisation process and has the structural formula:

$$\left[-CH_2 - \underset{CH_3OOC}{\overset{CH_3}{\underset{|}{\overset{|}{C}}}} - \right]_n$$

PMMA is a hard, rigid, thermoplastic polymer having exceptional optical clarity combined with good resistance to water, chemicals, alkalies and dilute acids, although it is attacked by concentrated acids and strong oxidising agents. Furthermore, PMMA exhibits excellent outdoor weathering resistance.

The resin has a very regular, linear structure and consequently tends to be crystalline with a high softening point ($125^{\circ}C$) and glass transition temperature ($105^{\circ}C$). PMMA can be rendered more flexible by copolymerising with other monomers such as ethyl acrylate and 2-ethylhexyl acrylate, whilst cross-linkage, which improves the chemical and physical properties, results in the formation of thermosetting resins. Thermosetting resins are also formed by co- or terpolymerisation with a variety of monomers including styrene, acrylonitrile and vinyltoluene, with cross-linking agents often being present to effect curing of the resin.

Thermoplastic resins are soluble in a variety of solvents, although the cross-linked materials are far less soluble. Coatings may be formed by solvent evaporation, that is, non-convertible coatings, although self-cross-linking binders (i.e. convertible coatings) are also used. Water-soluble and aqueous dispersions of acrylic resins are very popular as binders for surface coatings, e.g. emulsion paints for domestic use.

2.7 EPOXY RESINS

Epoxy resins are cross-linked polymers derived from reactions involving the epoxide or oxirane grouping, $>\!\!\overset{O}{\overset{\diagup\diagdown}{C\text{—}C}}\!\!<$. The principal raw material in resin preparation is bisphenol A which is formed by the reaction between phenol and acetone:

$$2\ \text{C}_6\text{H}_5\text{OH} + \text{CH}_3\text{COCH}_3 \longrightarrow \text{HO-C}_6\text{H}_4\text{-C(CH}_3)_2\text{-C}_6\text{H}_4\text{-OH}$$

Bisphenol A

Epoxy resins are produced by inter-reaction of bisphenol A and epichlorhydrin:

$$\text{HO-C}_6\text{H}_4\text{-C(CH}_3)_2\text{-C}_6\text{H}_4\text{-OH} + \underset{\text{Epichlorhydrin}}{\text{CH}_2\text{-CH-CH}_2\text{Cl}}$$

$$\underset{\text{Epoxy resin}}{\text{CH}_2\text{-CH-CH}_2-[\text{O-C}_6\text{H}_4\text{-C(CH}_3)_2\text{-C}_6\text{H}_4\text{-O-CH}_2\text{-CH(OH)-CH}_2]_n\text{-O-C}_6\text{H}_4\text{-C(CH}_3)_2\text{-C}_6\text{H}_4\text{-O-CH}_2\text{-CH-CH}_2}$$

$$+\ n\text{HCl}$$

These epoxy resins are also known as diglycidyl ethers of bisphenol A since the terminal groups, $\text{CH}_2\text{-CH-CH}_2\text{-}$, are glycidyl ether groups.

Liquid epoxy resins, having an average molecular weight of about 400, are prepared by reaction of bisphenol A and epichlorhydrin in molar ratios of 1:10 to 1:4 at 60°C in the presence of sodium hydroxide. Lower molar ratios (1:1.4 to 1:1.2) and higher temperatures (up to 100°C) yield solid epoxy resins having average molecular weights of 1500.

These epoxy resins cross-link relatively slowly, even at elevated temperatures, but the cyclic epoxide grouping can react with a variety of curing agents to open the ring, and polymers form by condensation.

Tertiary amines, R_3N where R may be an alkyl or an aryl group, as well as tertiary amine salts, are used as curing agents for epoxy resins. Typical tertiary amine curing agents are benzyldimethylamine and triethanolamine whilst salts such as the tri-2-ethylhexoate salt of tris (dimethyl aminomethyl) phenol are also used. Tertiary amine curing agents are sometimes referred to as catalysts since they induce direct linkage of the epoxide groups although the amounts required to effect reaction suggest their action is not purely catalytic. They do have an accelerating or catalytic effect, however, when used in conjunction with anhydride curing agents.

Tertiary amine salts are sometimes preferred since they permit greater additions of the curing agent to be added without affecting the pot life of the epoxy resin.

Polyfunctional amines such as diethylenetriamine and *m*-phenylenediamine are also used as curing agents for epoxy resins. In general, aliphatic amines effect faster cures and function at room temperature, whilst aromatic amines are less reactive but provide resins having higher T_g values.

Most amines, however, are somewhat toxic and may cause skin reactions or respiratory difficulties, although these toxic effects can be alleviated by using amine adducts or polyamides. Typical examples of such adducts are those produced by reaction of the diglycidyl ether of bisphenol A and diethylenetriamine and that from acrylonitrile and diethylenetriamine. Polyamides are produced by reacting amines with the fatty acids of vegetable oils.

Amine, amine adduct and polyamide-cured epoxy resins are known as cold-curing or room temperature curing materials since they do not require heat to effect cross-linking of the resin. Such materials are used as 'two pack' systems for paints, one component containing the resin and pigment whilst the curing agent is contained separately. The two components are then mixed just prior to paint application. Bisphenol A-epichlorhydrin resins are widely used in the surface coating field.

When epoxy resins are blended with other resins containing active groups, such as phenolic, amino and acrylic resins, the cross-linked copolymer exhibits the characteristics of both precursors. Epoxy resins may also be modified by esterification, that is, reaction of hydroxy and epoxy groups within the molecule with a carboxylic acid to yield an ester.

When epoxy resins are blended with phenolic resins, the mixture is cured by heat at 180-200°C and, during reaction, the hydroxy groups in the phenolic resin react with the epoxy groups to form methylol groups:

$$\underset{R}{\text{HO-C}_6\text{H}_3}-\text{OH} + \text{H}_2\text{C}\overset{O}{-}\text{CR}-\text{R}' \longrightarrow \underset{R}{\text{C}_6\text{H}_3}-\text{O}-\text{CH}_2-\overset{\text{OH}}{\text{CH}}-$$

These methylol groups, together with those present in the phenolic resin, react with hydroxy groups in the epoxy resin to yield a cross-linked polymer exhibiting excellent chemical and heat resistance.

Blends of epoxy and amino resins also require baking to effect cross-linkage, reaction occurring at 200°C, although the curing temperature may be reduced to 150°C by the use of acidic catalysts. The cross-linked copolymers are hard and exhibit good colour and thermal stability.

In general, higher molecular weight ($c.$ 4000) epoxies are blended with phenolic and amino resins, although lower molecular weight resins are blended with acrylic resins to yield thermosetting acrylic copolymers used for domestic appliances. Coal-tar is also often used to modify epoxy resins to both reduce the cost and increase the sea water resistance of epoxy resin coatings.

Esterification of epoxy resins (commonly of molecular weights of about 1500, i.e. $n = 4$ in the structural formula given earlier) with the fatty acids present in vegetable oils yields the epoxy esters. These epoxy esters form films which are similar to those formed by alkyd resins but possess superior chemical resistance, greater flexibility and improved adhesion. Drying, semi-drying and non-drying oils are used to prepare epoxy esters and, like oleoresinous varnishes and alkyds, these esters are classified on the basis of oil length. The superiority of epoxy ester films compared to alkyds is due to the absence of ester linkages within the backbone of the epoxy ester polymer.

The characteristics of films formed by epoxy esters are determined by the epoxy resin, the fatty acid and the oil length. The mechanical properties of the film are provided by the epoxy backbone whilst the durability and colour retention characteristics are determined by the

fatty acids present in the oil. Long drying oil and semi-drying oil epoxy esters are, with the incorporation of driers, air drying, although the films formed exhibit greater chemical resistance, notably in alkaline media, than the comparable alkyd films. Medium or short drying oil or semi-drying oil epoxy esters require stoving or baking for film formation and the films formed possess excellent mechanical properties and good adhesion.
Short semi-drying oil and non-drying oil epoxy esters are usually blended with amino resins to yield very hard films possessing excellent chemical resistance. These resins also require stoving for film formation to occur.

2.8 POLYURETHANE RESINS

Polyurethane resins are polymers containing the urethane grouping, $-\overset{H}{\underset{|}{N}}-\overset{O}{\underset{\|}{C}}-O-$, within the main polymer backbone and they are formed by the reaction of isocyanates, R-NCO, with hydroxyl compounds.

The isocyanate grouping is very reactive and will interact with water, alcohols, acids and bases as well as with ureas, urethanes, phenols and amides. These reactions are summarised in Table 2.2 and some of them are very important in the technology of polyurethane surface coatings.

Polyurethanes prepared for surface coatings are derived from di-isocyanates which form linear polymers upon reaction with diols (aliphatic glycols) and cross-linked or network polymers when reacted with polyols (e.g. polyesters, polyethers and some vegetable oils as well as water). The most commonly used di-isocyanate is toluene di-isocyanate (TDI) which exists in two forms, the 2,4 and the 2,6 isomers,

TDI is used as a mixture of the two isomers in the ratio of 80:20 or 65:35 of the 2,4 to 2,6 isomers. Other di-isocyanates having commercial importance are diphenyl-methane di-isocynate (MDI),

OCN–⟨phenyl⟩–CH₂–⟨phenyl⟩–NCO, and hexamethylene di-isocynate (HDI), $OCN-(CH_2)_6-NCO$.

	Reactant	Product	Decomposition product	Further reaction
(a)	Water	R·NH·COOH (a carbamic acid)	R·NH$_2$ (an amine)	As for (d)
(b)	Alcohol, R'·OH	R·NH·CO·OR' (a urethane)	—	—
(c)	Acid, R'·COOH	R·NH·CO$_2$·COR' (an anhydride)	R·CO·NH$_2$ (an amide)	As for (h)
(d)	Amine, R'·NH$_2$	R·NH·CO·NH·R' (a urea)	—	As for (e)
(e)	Urea, R'·NH·CO·NH·R"	R·NH·CO·NR'·CO·NH·R" (a biuret)	—	—
(f)	Urethane, R'·NH·CO·OR"	R·NH·CO·NR'·CO·OR" (an allophanate)	—	—
(g)	Phenol, C$_6$H$_5$·OH	R·NH·CO·OC$_6$H$_5$ (a urethane)	—	—
(h)	Amide, R'·CO·NHR"	R·NH·CO·NR'·COR" (an acylurea)	—	—

TABLE 2.2 Reaction of Isocyanates, R·NCO

There are two broad classifications of isocyanate-based coating systems, namely the one component or single pack system and the two component or two pack system. Each can form urethane resins in several different ways so that further subdivisions are possible.

2.8.1 Single Pack Systems

Single pack coating systems commonly form urethane resins by air curing (oxidative polymerisation), moisture curing and heat curing (stoving).

Air curing systems are known as polyurethane-alkyds or uralkyds (abbreviated form of urethane-alkyds) since they are similar to air drying alkyds except that a di-isocyanate, yielding urethane linkages, is used instead of a polycarboxylic acid, which yields ester linkages, as in alkyd resins. Urethane-alkyds are prepared in a similar way to alkyds using a two stage process. In the first stage, a monoglyceride or partial ester is prepared by the reaction of an oil with a polyhydric alcohol or polyol. After the monoglyceride is formed, a di-isocyanate, typically TDI, is added and reacts with hydroxyl groups in the partial ester (reaction (b) in Table 2.2) to yield the oil-modified polyurethane:

$$\sim\!\!O-\underset{O}{\overset{H}{C}}-\overset{H}{N}-\!\!\bigcirc\!\!-\overset{H}{N}-\underset{O}{\overset{H}{C}}-O-\underset{\underset{R_m}{|}}{\overset{CH_2-O-\overset{N}{\underset{O}{C}}-\overset{H}{N}-\!\!\bigcirc\!\!-\overset{H}{N}-\underset{O}{\overset{H}{C}}-O\sim}{\overset{|}{CH}}}$$

where Rm is the unsaturated fatty acid chain from the oil.

Urethane-alkyd coatings are characterised by good abrasion resistance and high resistance to water and mild corrosives. The type of oil and the oil length (the oil/resin ratio) used in the preparation of polyurethane-alkyds determines, in an analogous manner to conventional alkyd resins, the characteristics of the cured film. In general, the resistance to yellowing increases with the degree of unsaturation of the fatty acids in the oil. Similarly, film durability increases with the increase in oil length.

The curing mechanism of urethane-alkyds is similar to that for alkyds, that is, aerial oxidation of the unsaturated fatty acid residues followed by cross-linkage initiated by decomposition of the oxidation products. This oxidation of uralkyds is, like that of alkyd resins, too slow for practical applications in the absence of metallic driers, although the curing rate of the urethane-alkyds is faster than that of the alkyds.

In contrast to the air curing uralkyds, the prepolymers of moisture curing polyurethane systems are branched and isocyanate terminated. The prepolymers are derived from the reaction of excess TDI with polyols such that there is a low free di-isocyanate content (under 5%) but there is an excess of isocyanates in the reaction product. Upon exposure to atmospheric moisture, reactions (a) and (d) in Table 2.2 occur with liberation of carbon dioxide which diffuses out of the film. These systems form hard films possessing good flexibility and high chemical and abrasion resistance.

The stoving or heat cured systems are known as blocked adducts since the resin precursor is an adduct of TDI and a polyhydric alcohol 'blocked' or rendered inactive by reaction with phenol. Typically, excess TDI is reacted with a polyol such as trimethylol propane to form a urethane containing free isocyanate groups (reaction (b) in Table 2.2):

$$C_2H_5 - C(CH_2OH)_3 + 3\ OCN-\text{[Ar]}-CH_3\ (NCO) \longrightarrow$$

$$C_2H_5 - C\left[CH_2 - O - CO - NH-\text{[Ar]}-CH_3\ (NCO)\right]_3$$

The free isocyanate groups may then be reacted with a phenol (reaction (g) in Table 2.2) to form a non-reactive and therefore very stable blocked adduct:

$$C_2H_5 - C\left[CH_2 - O - CO - NH-\text{[Ar]}-CH_3\ (NCO)\right]_3 + 3\ \text{PhOH} \longrightarrow$$

$$C_2H_5 - C\left[CH_2 - O - CO - NH-\text{[Ar]}-CH_3\ (NH-CO-O-Ph)\right]_3$$

When, however, these blocked adducts are subjected to heat, typically 150°C, the free isocyanate groups are regenerated and cross-linkage can occur with added hydroxy compounds present in the binder formulation. Coating systems based on blocked adducts have the great advantage of stability during storage whilst the highly cross-linked polyurethane film exhibits excellent chemical and abrasion resistance.

2.8.2 Two Pack Systems

Two types of two pack polyurethane systems are used for surface coating applications, namely a cold curing isocyanate-polyol system and a catalyst curing isocyanate adduct-polyol system. Both two pack systems form films by *in situ* polymerisation and the two components of the coating formulation are mixed just prior to application.

The isocyanate-polyol system consists of an isocyanate adduct, such as the reaction product of TDI and trimethylol-propane or polymerised TDI (that is, a polycyanurate), so that the isocyanate component of the two pack system is rendered non-volatile. Alternatively, a low volatility isocyanate such as diphenylmethane di-isocyanate may be used. The low volatility is necessary to reduce the toxicity hazard associated with the use of isocyanates.

The polyol component may be a polyester or a polyether, which must contain free hydroxy groups, and urethane formation occurs through reaction (b) in Table 2.2. It should be noted that the isocyanate component will tend to react with any water present, namely, atmospheric water vapour or adsorbed water on the substrate, so that there may be a net excess of the hydroxy component in the final film. This leads to somewhat softer films with reduced chemical resistance. To offset this effect, a slight excess of the isocyanate component is normally included but if the isocyanate:polyol ratio is too high, brittle films may result although they will have enhanced chemical resistance.

The other two pack system has, for one component, an isocyanate-terminated prepolymer similar to those used in the moisture curing single pack system, whilst the second component is a catalyst, usually a tertiary amine. Film formation occurs through reactions (d) and (e) in Table 2.2 although a stoving regimen is frequently employed to accelerate the curing rate without the need for excessive catalyst levels. The films consist of fairly densely cross-linked polyurethanes and therefore exhibit excellent chemical resistance. This system has the double advantage

of long pot life of the mixed components whilst the
toxity of the prepolymer is considerably reduced due
to the low volatility of the isocyanate terminated
prepolymers.

Whilst the polyol components of the two pack systems
are oil-free, it is possible to incorporate certain
classes of oils and oil modified alkyds having a high
hydroxy group content as co-reactants. These additions
generally increase film flexibility and improve water
resistance.

The principal characteristics of polyurethane coatings
are their high chemical resistance, good durability, high
film strength and excellent adhesion to a wide variety
of substrates.

2.9 CHLORINATED RUBBERS

Natural rubber, which consists predominantly of
1,4-polyisoprene, $\left[CH_2 - \underset{\underset{CH_3}{|}}{C} = CH - CH_2 \right]_n$, is not used
in paint formulations as such but its chemical derivatives,
notably chlorinated rubber, are used fairly extensively.
Chlorinated rubber is prepared by chlorination of a
solution of masticated crepe rubber in carbon tetrachloride or chloroform at 80-100°C. This reaction yields
a stable material containing 60-65% combined chlorine,
e.g. the Aloprene materials marketed in the United
Kingdom by I.C.I. Ltd. Although the actual reactions
involved in the chlorination process are not fully
established, the first reaction product is thought to
be a cyclic compound:

$$\left[CH_2 - \underset{\underset{CH_3}{|}}{C} = CH_2 \right]_n + Cl_2 \longrightarrow \left[\begin{array}{c} -CH_2 - \overset{CH_3}{\underset{|}{C}} - \overset{Cl}{\underset{|}{CH}} \\ -CH_2 - \underset{|}{CH} \quad \underset{|}{CH_2} \\ \underset{|}{C} = CH \\ CH_3 \end{array} \right]_n$$

This initial cyclic compound may then undergo various
addition and substitution reactions to yield a complex
mixture of products. As a result, chlorinated rubbers
are available in different viscosity grades with the
lower viscosity materials being used for surface coatings.

Chlorinated rubbers are used for surface coating
applications as solutions in aromatic hydrocarbons when
film formation occurs by solvent evaporation, although

fairly large additions of plasticisers (see Section 4.1) are required to prevent excessive film brittleness. Chlorinated rubbers are also incorporated with other binders such as the alkyd resins, to increase the water and chemical resistance of the paint film. Chlorinated rubber coatings, when properly formulated, exhibit excellent resistance to most aqueous media, including acids and alkalis, whilst being virtually impermeable to water. Their principal use is in corrosion resistant coatings where exposure conditions are likely to be severe.

2.10 CELLULOSIC POLYMERS

Cellulose is the most widely occurring natural polymer, being the main constitutent of the cell wall of all plants. The primary sources of cellulose for a number of industrial applications are cotton fibre and wood, but there are a number of other sources.

Cellulose is a polyanhydroglucose:

$$\left[\begin{array}{c} H \\ HO-C \\ OH \\ C \\ H \\ CH_2OH \end{array} \right]_n \quad \text{or, more simply} \quad \left[\begin{array}{c} HO \\ OH \\ CH_2OH \\ O \end{array} \right]_n$$

The molecular weight of the polymer (that is, the degree of polymerisation) varies with the source of the cellulose. Whilst cellulose itself is insoluble, many of its derivatives are soluble and these compounds are utilised in the preparation of cellulosic lacquers (amongst a wide variety of other applications). The cellulose nitrate lacquers were once of great importance for the preparation of fast drying lacquers for the automobile industry but they have been largely superseded by modern synthetic resin binders.

Cellulose nitrate, more commonly known as nitrocellulose, is prepared by the action of a mixture of nitric and sulphuric acids on cellulose at 30-40°C to yield a viscous product, possibly:

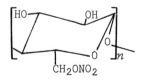

This product is generally degraded by the action of water at temperatures up to 160°C under pressure to yield a less viscous material containing 11-12% nitrogen. The degree of reaction (that is, esterification) between cellulose and nitric acid can be regulated by control of the reaction conditions. Cellulose may be esterified to yield a product with a higher nitrogen content (12.4-13.5%) which corresponds to a di-nitrate and is commonly referred to as nitrocellulose. This initial product is dehydrated with ethanol under pressure to yield cellulose nitrate containing about 5% water and 30% alcohol.
Cellulose nitrate is only used commercially in surface coatings as a plasticised material, the plasticiser being required to provide film flexibility and adhesion. The material is soluble in a wide range of solvents and the solutions can tolerate additions of both aromatic and aliphatic hydrocarbon diluents. Film formation is by solvent evaporation and the cellulose nitrate films are resistant to water and dilute acids but are decomposed by alkalies and concentrated acids. For completeness, it should be mentioned that celluloid, one of the first commercially exploited synthetic polymers, is cellulose nitrate plasticised with camphor.

2.11 PHENOLIC RESINS

When a phenol reacts with an aldehyde, usually formaldehyde, in the presence of a catalyst, low molecular weight resins are formed. Resol resins are formed by the reaction of phenol with molar excess formaldehyde under alkaline conditions. The initial reaction products for phenol are *ortho*- and *para*-methylolphenols:

[structures: ortho-hydroxybenzyl alcohol (OH, CH_2OH) and para-hydroxybenzyl alcohol (OH, CH_2OH)]

These products undergo further reaction with formaldehyde to form di- and trimethylolphenols. Upon heating, these methylolphenols can undergo self-condensation to yield polynuclear phenols, network polymers with methylene, $-CH_2-$, links between the nuclei. Novolak resins are formed by the reaction between molar excess of phenol and formaldehyde under acidic conditions. This reaction yields three different dihydroxydiphenylmethane isomers, the 2,2' isomer being:

HO—C₆H₄—CH₂—C₆H₄—OH

(structure: two ortho-hydroxyphenyl groups linked by CH₂)

The dihydroxydiphenylmethanes can undergo further reaction with formaldehyde to yield polynuclear phenols. These low molecular weight novolak resins may be cross-linked to form network polymers by heating in the presence of a cross-linking agent such as hexamethylene-tetramine.

Resol and novolak resins are little used in surface coatings although when modified, these phenolic resins have importance. When heated with excess rosin, resols form complex acids which may be esterified with a polyol, typically glycerol. For varnishes, the ester is then heated with a drying oil such as linseed and tung oil. Coatings based on short oil combinations are cured by stoving, whilst long oil combinations are air-drying. Rosin modified phenolic resin-drying oil combinations form coatings exhibiting good chemical resistance, although they tend to be somewhat brittle.

Alternatively, phenolic resins can be rendered oil-soluble by the use of certain substituted phenols, for example, *para*-tertiary-butyl phenol. Varnishes prepared from these so-called 100% phenolic resins have an inherently higher durability than their rosin-modified counterparts.

2.12 OTHER BINDER MATERIALS

2.12.1 Silicone Resins

Silicone resins are polymers with a backbone comprised of alternate silicon and oxygen atoms, the oxygen atoms linking the alternating silicon atoms to which organic groups are also attached. Compounds having the general formula $R_3Si-(OSiR_2)_n-OSiR_3$, where R may be H, Cl or an organic grouping, have the generic name siloxanes, but are commonly known as silicones.

There are three fundamental units containing silicon-oxygen pairs in organosiloxane compounds, namely the monofunctional R_3SiO-(M) group, the difunctional R_2SiO-(D) group and the trifunctional $RSiO_3$-(T) group, and in all three, the oxygen atom is 'shared' with another silicon atom. The R grouping is usually an alkyl group such as methyl, CH_3-, or the phenyl group, C_6H_5-.

There are in fact three types of silicone polymer: silicone rubbers, silicone fluids or oils and silicone resins. Only the last named is of direct interest for surface coatings but brief mention of the other silicones will be made for completeness.

Silicone polymers (siloxanes) are formed when di- or trichlorosilanes (R_2SiCl_2 or $RSiCl_3$ where R is typically CH_3- or C_6H_5-) are hydrolysed to form silanols ($R_2Si(OH)_2$ or R_3SiOH) which, being unstable, then spontaneously undergo condensation to form the siloxane. Both the chlorosilane used and the hydrolysis conditions influence the resultant siloxane. In general, the difunctional dichlorosilanes form linear polymers, that is, they have a linear chain of D units terminated by M units whilst the trifunctional trichlorosilanes form cross-linked structures containing a high proportion of T units:

$$Cl-\underset{R}{\overset{R}{Si}}-Cl \xrightarrow{H_2O} HO-\underset{R}{\overset{R}{Si}}-OH$$

$$HO-\underset{R}{\overset{R}{Si}}-\boxed{OH + H}O-\underset{R}{\overset{R}{Si}}-\boxed{OH + H}O-\underset{R}{\overset{R}{Si}}-OH \xrightarrow{-H_2O} \left[\underset{R}{\overset{R}{Si}}-O\right]_n$$

D units

$$Cl-\underset{Cl}{\overset{R}{Si}}-Cl \xrightarrow{H_2O} HO-\underset{OH}{\overset{R}{Si}}-OH \longrightarrow \begin{array}{c} \sim Si-O-Si\sim \\ || \\ OR \\ || \\ \sim Si-O-Si\sim \\ || \end{array}$$

T units

Hydrolysis of a dichlorosilane, such as dimethyldichlorosilane, with water yields a mixture of the linear polydimethylsiloxanediol, with polydimethylcyclosiloxane

$$HO-\underset{CH_3}{\overset{CH_3}{Si}}-O-\underset{CH_3}{\overset{CH_3}{Si}}-O-\underset{CH_3}{\overset{CH_3}{Si}}-OH$$

Polydimethylsiloxanediol

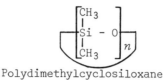

Polydimethylcyclosiloxane

whilst hydrolysis in sulphuric acid yields predominantly the linear polymer. It is also possible to hydrolyse the dichlorodimethylsiloxane under conditions such that the cyclic siloxane is the major product. Hydrolysis of methyltrichlorosilane yields highly cross-linked polymers containing the repeating unit:

$$-O-\underset{\underset{|}{O}}{\underset{|}{Si}}-O-$$
$$\overset{|}{CH_3}$$

The condensation reactions of silanols are catalysed by both acids and bases and also by metallic salts of organic acids, notably tin salts.

Silicone rubbers (or, more correctly, elastomers) are high molecular weight linear polymers which are cross-linked by suitable catalysts to obtain the desired elastomeric properties. Silicone fluids are also linear polymers but they have lower molecular weights than the elastomers and are not cross-linked. Both the elastomers and fluids are derived from hydrolysis of dichlorosilanes.

In contrast, silicone resins are cross-linked network polymers derived from the hydrolysis of trichlorosilanes which contain a proportion of dichlorosilanes in the reaction mix to moderate the degree of cross-linking. The resins used for surface coatings generally contain both methyl and phenyl radicals, since a high proportion of methyl groups yields hard but mechanically weak resins whilst a preponderance of phenyl groups gives stronger but more brittle resins. Consequently, mixtures of methyltrichlorosilane and phenyltrichlorosilane together with their dichlorosilane counterparts are hydrolysed to yield resins of suitable properties.

The final resin consists predominantly of a mixture of silanol terminated linear and cross-linked polymers, and is soluble in various solvents. This partially polymerised material is fully cross-linked by heating, in the presence of a catalyst, to condense the remaining silanol groups. These pure silicone resins are only used in baking finishes but they exhibit excellent chemical and water resistance together with non-stick properties.

Silicone resins may also be blended or copolymerised with other resins such as alkyds and polyurethanes to increase their durability, temperature resistance and water repellancy. In such operations, a low molecular weight silanol terminated siloxane is heated with an alkyd to yield a copolymer.

2.12.2 Bituminous Materials

There are a number of materials broadly referred to as bituminous, namely bitumens, asphalts, tar and pitch, although these materials in fact differ quite markedly. Furthermore, substances referred to as asphalt in the United States of America are known as bitumen in the United Kingdom, whilst asphalt in the UK is a mixture of bitumen and finely divided mineral matter. In broad terms, bitumens are crude (or petroleum) oil residues, either occurring naturally or obtained by distillation (although the former are also sometimes known as asphalts). Tar is obtained by the destructive distillation of coal and other carbonaceous materials. The residue from distillation of tar is known as pitch.

Several grades of bitumen are produced by variation of the distillation process conditions and these grades vary in properties and colour, depending upon the source and the process conditions. Bitumens used in bituminous paints are the naturally occurring products or the so-called asphaltic bitumens obtained by distillation of crude oils from the Middle East as well as North and South America. These natural bitumens have high softening points (up to 150°C) and the distillation products are often oxidised by having air blown through them in order to raise their softening points.

Bitumens dissolved in hydrocarbon solvents are often used to form surface coatings by solvent evaporation. These coatings provide good water and chemical resistance since bitumens are resistant to moderate concentrations of non-oxidising acids and alkalies at ordinary temperatures Bitumens are also blended with epoxy resins to yield the so-called coal tar epoxy coatings in which the water impermeability and chemical resistance of the bitumen reinforces the corrosion and solvent resistance as well as the good mechanical properties of the epoxy resin coatings. Bitumens may also be blended with drying oils

such as linseed and tung oil to yield coatings exhibiting
good durability and gloss retention as well as improved
weathering resistance.

2.12.3 Inorganic Silicates

Inorganic silicates are used in conjunction with zinc
dust to produce very hard, abrasion, chemical and
oxidation resistant coatings which bond extremely well
to shot blast cleaned steel. These inorganic silicates
are formed by fusing a mixture of sand or silica with
sodium and/or potassium carbonate and then digesting the
fused or vitrified mass in water. When zinc dust and
zinc oxide or calcium oxide are added and the mixture
is applied to a surface, evaporation of the water and
the action of atmospheric carbon dioxide forms a
silicic acid gel which then reacts with the zinc oxide
or calcium oxide (and, in part, the zinc dust present)
to form insoluble silicates. Various additives, such
as lithium silicate and potassium phosphates, are often
included to accelerate the curing rate of these two pack
'self-curing' formulations.

The zinc dust-inorganic silicate coatings are extremely
durable and provide exceptional protection against
corrosion whilst being unaffected by temperatures of
up to $c.$ 500°C. These materials do not dry well under
wet conditions and are attacked by acids and strong
alkalies; they are unaffected by sea water and form
thick coatings (greater than 100 μm) in a single application. The zinc dust-inorganic silicate coatings can
be used as the sole means of protection against
corrosion for marine structures although top coats may
be provided for aesthetic reasons.

Organic-inorganic silicates have also been developed
for use in zinc rich primers. In these formulations
zinc dust is mixed with ethyl orthosilicate as a binder.
Ethyl orthosilicate is a mixture of condensed
tetraethoxysilanes which undergo further hydrolysis in
air to liberate ethyl alcohol leaving a zinc silicate,
wholly inorganic, surface coating, which has superior
properties to coatings derived from sodium silicate.
More recently, single pack ethyl orthosilicate-zinc
dust coatings have been developed in which the ethyl
orthosilicate is less hydrolysed than in the earlier
two pack formulations so that longer pot lives (up to
one year)are possible.

2.13 BIBLIOGRAPHY

Fibres, Films, Plastics and Rubbers - Handbook of Common Polymers, W.J. Roff and J.R. Scott, Butterworths, London (1971).

Organic Polymer Chemistry, K.J. Saunders, Chapman and Hall, London (1973).

Introduction to Paint Chemistry, G.P.A. Turner, Blackwells, Oxford (1967).

Technology of Paints, Varnishes and Lacquers, C.R. Martens, Reinhold, New York (1968).

3 Pigmentation of paints

Paints and allied materials are composed of resinous materials, pigments, certain minor additives and, generally, solvents. Within these broad groupings, however, there are numerous different materials, each imparting special properties to the paints into which they are incorporated.

The theory underlying the use of resins and pigments in paints has been discussed in Chapter 1. In this chapter the 'working' characteristics of the various pigments and extenders used in paint formulations will be described. In view of the very large number of products that can be incorporated into paints, only the most important types used in modern coating systems will be described here. A more comprehensive treatment of paint pigmentation can be obtained from the books listed in the bibliography at the end of this chapter.

3.1   PIGMENTS, EXTENDERS AND DYES

A pigment, which can be organic or inorganic in origin, may be defined as a solid material, in the form of small discrete particles, which is incorporated into, but remains insoluble in, the paint medium. A pigment confers a number of attributes to a paint film, notably colour and opacity, whilst influencing the degree of resistance of the film to light, contaminants and other environmental factors, as well as modifying the flow properties of the liquid paint.

There is a further class of paint additives that are also insoluble in the paint medium but which impart little or no opacity or colour to the film into which they are incorporated. These materials are known as extenders and they are all of inorganic origin. Extenders are incorporated into paints to modify the flow properties, gloss, surface topography and the mechanical and permeability characteristics of the film.

Solids with pigmentary value, but which are soluble in the paint medium, are also used in paints and these are termed dyes or dye-stuffs. However, it is impossible to classify a material as a dye or as a pigment without knowledge of its solubility in the medium in which it is to be used and, in certain instances, it is possible for a given material to be considered either as a pigment or a dye according to the paint medium.

Dyes are exclusively of organic origin and generally, although not necessarily, dyes (and organic pigments) are only incorporated into paints whose prime function is decorative rather than protective.

The categorisation of pigments and extenders by chemical class is shown in Table 3.1. Certain broad generalisations can be drawn concerning the properties of the pigments falling within each class.

Pigments within the natural inorganic class, that is pigments of mineral origin that are mined in the chemical form in which they are ultimately used, are generally coarse textured and dull in colour. The light stability of this class of pigment is excellent and they are capable of providing a high degree of opacity to films in which they are incorporated. Furthermore, the chemical resistance of this class of pigment is normally very high.

Synthetic inorganic pigments are chemically prepared from inorganic raw materials and the chemical and physical composition of the derived pigment is dissimilar to that of the parent materials. The texture of synthetic inorganic pigments is much finer than that of the naturally occurring inorganic pigments and this renders them more readily dispersable during paint preparation. Both the stability of this class of pigment to light and the opacity in paint films is good. The colour range is large, ranging from dull, drab colours to clean and intense colours. The chemical resistance of these pigments, however, is somewhat variable, depending on their chemical composition. For example, Ultramarine blue is highly stable towards alkalies but is sensitive to acids, whereas Prussian blue is stable to acids but unstable in the presence of alkalies.

The naturally occurring pigments or dyes of organic origin are usually derived from animal and vegetable products and, in general, they are not stable towards

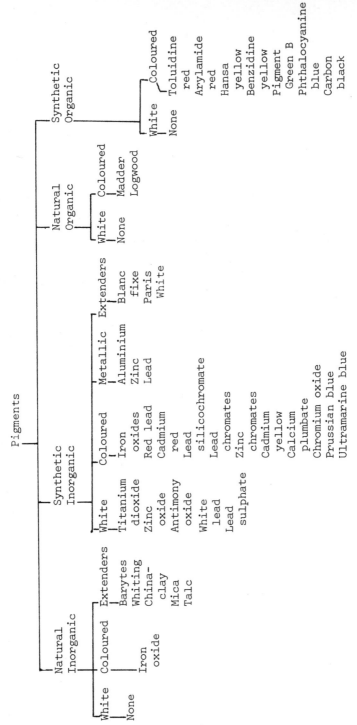

TABLE 3.1 Classification of Pigments

light. As a class, they are mainly of historical
interest and although certain types are still used for
specialist outlets, such as artist's colours, they will
not be discussed in this chapter.
 Synthetic organic pigments and dyes, in common with
the synthetic inorganic pigments, have chemical and
physical properties that often differ markedly from
those of the constituent raw materials. Synthetic
organic pigments are very finely textured and they provide
clean, intense colours although, at the low rates of
addition normal with this class of pigment, they do not
provide a high level of opacity. Both the stability
towards light and the chemical resistance of synthetic
organic pigments tend to be variable, depending upon
the chemical composition of the pigment.
 For clarity, the various pigments used in paints will
be classified in the following sections under the headings
of inorganic pigments, organic pigments, and extender
materials.

3.2 INORGANIC PIGMENTS

The pigments within this class can conveniently be classified by colour. The extenders, although generally white
in colour, will be discussed separately.

3.2.1   White Pigments

Titanium Dioxide TiO_2

Titanium dioxide is a synthetic inorganic pigment widely
used, almost to the exclusion of other white pigments,
in many types of decorative and industrial paints and
allied materials. Titanium dioxide is non-toxic and it
is the most stable white pigment known. This stability
renders it suitable for use with all paint media, and for
exposure to all types of service environment.
 Titanium dioxide is produced in two crystallographic
forms, rutile and anatase, which differ in their chemical
structures and hence in their physical and chemical
properties. The rutile crystal has a more compact
structure than that of anatase and this results in rutile
titanium dioxide having a higher refractive index, higher
density and greater stability. Rutile grades
are generally favoured in the paint industry because of
their greater durability.
 Titanium dioxide pigment is prepared in two ways,
namely by the sulphate or the chloride routes. In the
sulphate process, the ore ilmenite, $FeO.TiO_2$, is dissolved

in sulphuric acid and the resultant solution of titanium sulphate is hydrolysed by boiling to produce a hydrated oxide, the iron remaining in solution. This oxide is then calcined at approximately 1000°C to form the pigmentary material titanium dioxide TiO_2. During the calcination process, the TiO_2 crystals are allowed to 'grow' both to the required crystal type and to the particle size which produces maximum opacity in paint media, $c.$ 0.25 μm.

The chloride process is the more modern production process, although manufacturing aspects make it the more technically difficult of the two routes. With this process, mineral rutile, an impure form of TiO_2, is reacted with chlorine to form the liquid titanium tetrachloride, $TiCl_4$, and this, after purification, is oxidised under carefully controlled conditions to produce titanium dioxide. By adjusting the oxidation conditions, both the crystal form and the particle size of the pigment can be controlled.

To aid subsequent dispersion in paint media and to control chalking (see below), titanium dioxide pigment is normally coated after manufacture with various inorganic or organic compounds. Because of this treatment the actual percentage of TiO_2 in any given grade of titanium dioxide pigment can vary but is typically in the region of 94-98%. Furthermore, the grade of pigment selected for any one particular application must be selected with some care, for example, a grade of titanium dioxide treated so that it is suited for use in emulsion resins would not necessarily exhibit similar performance characteristics in solvent soluble resins.

One of the most important attributes of titanium dioxide pigment is its high refractive index (i.e. rutile 2.7 and anatase 2.5) and this enables it to provide a degree of opacity in paint films unsurpassed by any other white pigment. The stability of the pigment is also an important factor and titanium dioxide can be used with suitably resistant polymers in paints exposed to contact with organic and inorganic acids and alkalies. The high melting point of titanium dioxide, $c.$ 1825°C, also renders it suitable for use in temperature resistant coatings.

An important feature of paints containing the rutile form of titanium dioxide is the ability of the pigment to control film degradation. This is attributable to its photochemical inertness which protects the film by shielding it from the action of light rays, the shielding efficacy deriving from the light scattering power of the pigment.

The anatase form of titanium dioxide, however, is photochemically active, that is it can accelerate breakdown of a paint film by converting light energy to chemical energy, leading to a progressive destructive oxidation of the resin component of the film. This loss of surface binder results in the pigment at the surface of the film being only loosely bound such that this layer can be removed by normal weathering processes. This phenomenon is known as chalking and, in general, films with a high chalking rate do not fail by cracking. However, the loss of film thickness that can occur with paints formulated solely on the anatase form of titanium dioxide is quite rapid, and in service such paints suffer from rapid loss of gloss and possibly colour. Furthermore, white staining of areas immediately below chalking paintwork can occur and such paints will also require regular maintenance to preserve an adequate film thickness and hence substrate protection. Anatase titanium dioxide is generally used as the principal white pigments only in exterior paints where a degree of chalking is required, for example, to produce a self-cleaning effect and to offset normal dirt deposition. More commonly, however, the anatase form is used in admixture with rutile titanium dioxide in order to control the chalking process.

The rutile form of titanium dioxide is inherently more stable than the anatase form, although paints pigmented solely with the rutile form will eventually exhibit a degree of chalking. The degree of chalking, however, is very much less than that observed with the anatase grade.

Titanium dioxide can be used in all types of paints and paint systems where the inclusion of white pigment is required. The inertness and stability of these pigments enable correctly formulated paints to withstand the severest of exposure conditions. Because titanium dioxide is non-toxic, it can be used with safety in coatings for the food packaging industries, in food factories and dairies and in paints for children's toys.

Zinc Oxide, ZnO

Zinc oxide is a synthetic inorganic white pigment prepared by vaporising metallic zinc at a temperature of $c.$ 900°C in the presence of oxygen. Zinc oxide is collected by condensing this vapour, and the particle size and shape of the pigment particles is determined by the cooling rate of the oxide after its formation. Rapid cooling produces a pigment in the form of small nodules, whilst slow cooling produces needle shaped crystals; when prepared in the latter form, the pigment is known as acicular zinc oxide.

As a pigment, zinc oxide is basic in nature and this results in interaction with certain types of paint resins, in particular those with a high acid value. This leads to the formation of zinc soaps within the structure of the film and can result in mechanical reinforcement of the strength characteristics of the paint film. However, such films normally become brittle rather quickly on exterior exposure. This may lead to premature failure. Consequently, zinc oxide is rarely used as the sole pigment in modern coatings, although it finds some use in admixture with other pigments. One important property of zinc oxide is its ability to inhibit mould growth, although to be effective, it must constitute at least 30% by weight of the film.

Antimony Oxide, Sb_2O_3

Antimony oxide is a synthetic inorganic white pigment widely used in the preparation of fire retardant paint systems. The pigment is prepared from metallic antimony using a similar technique to that used for the preparation of zinc oxide. The resulting antimony oxide pigment is non-reactive in paint media and has good obliterating powers.

Prior to the introduction of the rutile form of titanium dioxide, antimony oxide was used to modify the heavy chalking characteristics of the anatase form of titanium dioxide. However, the prime use of antimony oxide in modern coatings is, in conjunction with chlorine containing resins, in the formulation of fire retardant coatings. On exposure to fire, the chlorine gas liberated by decomposition of the resin component of the paint film reacts with the antimony oxide to produce a vapour of antimony chloride which blankets the flames.

White Lead, $2PbCO_3.Pb(OH)_2$

White lead is a synthetic inorganic pigment with a chemical composition approximating to a basic carbonate of lead, $2PbCO_3.Pb(OH)_2$. White lead, a traditional pigment, is prepared by a complex process involving exposure of sheets of metallic lead to an atmosphere comprising acetic acid vapour, water vapour, air and carbon dioxide. The water and acid attack the lead to produce a basic acetate which is converted into the carbonate by the action of carbon dioxide. Complete reaction of the lead to white lead is a lengthy process which can take up to several weeks.

As white lead is a basic material, it can react with resins of high acid value, typically oils and oil containing resins. This reaction results in the formation of lead soaps which, due to their fibrous texture, impart a degree of elasticity to the film. Consequently, this pigment forms a suitable basis for primers for use on substrates prone to movement, such as timber. The chalking rate of white lead containing films is, however, rather high and the flow, gloss and colour stability of such paints are inadequate to render them suitable for use as finishing coats.

One serious drawback to the continued use of white lead pigment is its high toxicity and this severely limits its acceptibility in modern paint coatings.

Basic Lead Sulphate, $2PbSO_4.PbO$

Basic lead sulphate is a synthetic inorganic white pigment prepared by treating lead monoxide, PbO, with sulphuric acid. The pigment is formed by precipitation and has a composition approximating to $2PbSO_4.PbO$.

Basic lead sulphate has superior flow properties in paint media to those of white lead (basic lead carbonate) and, consequently, it can be used in finishing paint coats. However, paint films containing both white lead and basic lead sulphate are discoloured in atmospheres contaminated with sulphur compounds and this, together with the known toxicity factor, limits the use of basic lead sulphate pigments in many applications.

The prime use of basic lead sulphate is in anti-corrosive coating systems for steelwork, particularly in marine applications, where it is normally used in conjunction with other pigments, such as aluminium and micaceous iron oxide. The mechanism of protection of basic lead sulphate containing paints is attributable to a reduction (polarisation) of the rate of the electrochemical reactions occurring at the corroding ferrous substrate.

3.2.2 Red Pigments

Red Iron Oxide, Fe_2O_3

Red iron oxide is an inorganic pigment of either natural or synthetic origin. Natural red oxides of iron are mined either as the mineral haematite, Fe_2O_3, or as haematite in its hydrated form. Generally, the characteristic colour of the pigment is developed by high temperature calcination. The presence of impurities adulterates the material such that the Fe_2O_3 content of the pigment is variable and can be as low as 50% in some instances. This variation in composition results in the red iron oxides having a wide range of shades and colours which extend from orange-red to deep brown. The colouring and opacifying strengths of the pigment are also dependent on the pigment composition and, generally, pigments with a low Fe_2O_3 content are inferior in these respects to the higher Fe_2O_3 containing grades. Natural red iron oxides are used as colourants in stains for timber, low cost metal primers and in fillers, cement and concrete.

Synthetic red iron oxides are softer pigments than the natural forms and, due to their greater purities (typically c. 99% Fe_2O_3), the synthetic materials are also cleaner and brighter in colour. Furthermore, the colour strength is greater. The synthetic red iron oxides can be prepared by two basic techniques, one involving the oxidation of ferrous carbonate obtained by the reaction of ferrous sulphate and sodium carbonate solutions. The second preparatory technique involves calcining ferrous sulphate in the presence of air at high temperature.

As with the natural forms, the colour and strength of the synthetic pigments can be altered by appropriate changes in the manufacturing process. Synthetic red iron oxides have a similar range of applications to that of the natural forms. However, their greater strength and cleanness of colour and somewhat softer texture permits their use in a wider variety of decorative and industrial paint systems.

Irrespective of derivation, red iron oxide pigments are characterised by a high resistance to alkalies and organic acids. However, they are slowly attacked and discoloured by mineral acids and can alter in shade at high temperatures. As a class, they are also efficient ultraviolet light absorbers and hence they can reduce photochemical degradation of the resin component of paint films during exterior exposure.

Red Lead, $PbO_2.2PbO$

Red lead is an orange coloured synthetic inorganic pigment used mainly as a protective pigment in primers for steelwork rather than as a colouring pigment. The pigment is prepared by heating lead monoxide, PbO, in air and the composition approximates to $PbO_2.2PbO$ containing some residual or unreacted PbO. The residual PbO content of the pigment largely determines its applications.

Red leads with a high proportion of free PbO rapidly react with linseed oil to form a solid substance and, consequently, this grade is unsuitable for use in paints. However, it is used in jointing mastics for plumbing and similar applications. Red leads with a low free PbO are non-setting when used in admixture with linseed oil so that this grade can be used in paints. Red leads are basic in nature and react with media of high acid values leading to the formation of lead soaps. The presence of such soaps leads to an improvement in the mechanical properties of paint films, but it also results in a progressive thickening of the liquid paint during storage.

Red leads darken on exposure to sulphur containing atmospheres and this, together with the poor flow properties exhibited by red lead containing paints, renders them unsuitable for use in finishing systems. The main use of red lead is in combination with linseed oil where it forms the basis of a range of highly effective protective primers for ferrous metals. However, the toxic nature of this pigment limits its use in modern coating systems.

Cadmium Reds, CdS-CdSe

A range of synthetic inorganic red pigments ranging in shade from orange to deep maroon can be obtained from compounds containing various ratios of cadmium sulphide, CdS and cadmium selenide, CdSe. The pigment is typically prepared by passing hydrogen sulphide gas through solutions of cadmium sulphate and sodium selenide and the pigment is obtained as a precipitate. A high temperature calcination process is then used to develop the final shade of the pigment.

The ratio of CdS to CdSe determines the final shade of the pigment, although 3CdS.2CdSe could be considered a typical composition. The cadmium colours are stable, non-toxic pigments which are fast to light and alkalies, and are stable up to $c.$ 500°C. They are decomposed, however, by acids. Cadmium pigments are used as colourants in decorative finishes for all substrates, particularly where alkali resistance is a requirement. Their temperature stability also enables them to be used in stoving cured paints and in paints subject to high ambient or service temperatures.

Basic Lead Silicochromate, $PbO:CrO_3:SiO_2$

Lead silicochromate is a synthetic inorganic pigment with a dull orange colour. This pigment is used in protective coatings for steelwork and its low tinting strength enables it to be used, in conjunction with suitable colouring pigments, at all levels in the coating system to provide corrosion defence 'in-depth'. The pigment is prepared by calcining lead chromate and lead silicate onto an inert silica core which yields a compound of complex composition, approximating to 45% PbO, 5% CrO_3 and 50% SiO_2.

The corrosion inhibiting capability of the pigment is thought to derive from both formation of inhibitive lead compounds (typically soaps in paint media with high acid value) and the release of inhibitive chromate ions in the presence of water. Lead silicochromate is a more versatile pigment than red lead due to its ability to be used with a wide variety of paint media to produce coatings of good can stability and flow. Despite the presence of a high proportion of lead compounds, lead silicochromate is less toxic than red lead.

3.2.3 Yellow Pigments

Lead Chromates, $PbCrO_4$

The lead chromates (or, commonly, lead chromes) comprise a number of types of synthetic inorganic pigments, ranging in colour from pale yellow through deep orange to scarlet. These chromates are an important class of pigment and they are used to impart colour and opacity to a wide range of decorative and industrial undercoats and finishing systems. They are characterised by good light fastness, high tinting strength and opacity, although, in common with other lead-based pigments, they are toxic in nature. Furthermore, lead chromate pigments are sensitive to atmospheric contaminants, being darkened by hydrogen sulphide and fading in the presence of sulphur dioxide. Despite the good light fastness of this class of pigments, bleaching by sulphur dioxide results in a gradual loss of colour in films containing lead chromates on prolonged exposure in industrial atmospheres. The colour of lead chromes is also adversely affected by alkaline conditions and, consequently, such paints should not be used on alkaline substrates such as plaster and concrete.

The pigments within the broad class of lead chromate pigments derive their colour from the presence of chromium containing anions. Normal lead chromate, more commonly known as mid chrome, $PbCrO_4$, is a bright yellow pigment obtained as a precipitate from the reaction between sodium chromate and lead nitrate solutions. The colour of the pigment can be varied in severals ways, for example, combining lead sulphate with the lead chromate produces paler shades, the actual colour being dependent upon the relative ratio of the two compounds. Conversely, treating lead chromate with alkali produces a range of deeper, redder shade yellows. Further, substituting part of the chromium content of lead chromate with a soluble salt of molybdenum produces bright scarlet shades, whilst substituting with a soluble salt of aluminium produces the so-called primrose chrome.

The range of lead chromate pigments can be classified into the following groups:

Primrose chrome,		$PbCrO_4 \cdot PbSO_4 \cdot Al_2(OH)_6$	Very pale yellow
Lemon chrome,	Pale,	$PbCrO_4 \cdot PbSO_4$	
	Normal,	$2PbCrO_4 \cdot PbSO_4$	
	Deep,	$3PbCrO_4 \cdot PbSO_4$	Becoming redder in shade.
Mid chrome,		$PbCrO_4$	
Orange chrome,		$PbCrO_4 \cdot Pb(OH)_2$	Deepest shade orange
		$2PbCrO_4 \cdot Pb(OH)_2$	
		$3PbCrO_4 \cdot Pb(OH)_2$	Palest shade orange
Scarlet chrome,		$PbCrO_4 \cdot PbMoO_4 \cdot PbSO_4$	

Zinc Chromates, $ZnCrO_4$

The zinc chromates are a class of synthetic inorganic pigments having a pale yellow colour, three forms of which can be used in paints.

Zinc chromate can be used as a colouring pigment in decorative paints. It is characterised by excellent light fastness, but its use is restricted due to poor opacity. The pigment has the advantage of being non-toxic and, furthermore, its colour is not altered by either exposure to sulphur containing atmospheres or to weak alkalies. The colour stability of zinc chromate when exposed to lime permits its use as a pigment in paints for plaster and concrete. Zinc chromates, however, are dissolved by acid and strong alkalies and they are slightly water soluble. The zinc chromates as a class are slightly basic in nature and they are therefore reactive with paint media of high acid value. This reactivity results in thickening during storage of the liquid paint.

Zinc chromate for colouring purposes is typically prepared by reacting zinc oxide with potassium dichromate and hydrochloric acid and the derived pigment has the composition $4ZnCrO_4 \cdot K_2O \cdot H_2O$. There tends to be, however, a low residual chloride content in the pigment. This residual chloride content restricts the use of this grade of zinc chromate in anti-corrosive paints.

Zinc chromate for use in anti-corrosive primers is prepared by either washing the grade prepared for pigmentary purposes free of chloride, or by using an alternative preparative route which does not involve the use of hydrochloric acid. Typically, the latter process involves the use of chromic acid in place of hydrochloric acid and the pigment has the composition $K_2CrO_4 \cdot 3ZnCrO_4 \cdot Zn(OH)_2$.

A third form of zinc chromate, know as zinc tetroxychromate, is exclusively used in the preparation of metal pre-treatment primers, alternatively known as etch or wash primers. These pre-treatment primers are widely used on ferrous and, particularly, on light alloy substrates, where they impart a high level of adhesion for subsequently applied coating systems. Zinc tetroxychromate is favoured as the protective, and pigmentary, pigment in these pre-treatments, as its lower solubility compared with zinc (potassium) chromate has been found to improve performance. Zinc tetroxychromate is prepared from zinc oxide and chromic acid and the composition approximates to $ZnCrO_4 \cdot 4Zn(OH)_2$.

In general, zinc chromates exhibit a slight solubility in water, with a consequent release of chromate ions, and, the latter have pronounced inhibitive properties towards ferrous and aluminium substrates. Consequently, zinc chromates are widely used as pigments in metal primers, provided they are free from residual chloride.

Yellow Iron Oxides, $Fe_2O_3 \cdot xH_2O$

Yellow iron oxides are inorganic pigments of either natural or synthetic origin. The naturally occurring forms are composed of hydrated iron oxides and they range in shade from a dull but clean yellow to a dark yellow-brown. A range of synthetic yellow oxides are also prepared and these are available in a wider range of shades than the naturally occurring varieties.

Yellow iron oxides possess the same general characteristics as the red oxides described above and are used in similar applications.

Cadmium Yellow, CdS

Cadmium yellow pigments are synthetic inorganic pigments varying in shade from primrose to orange. These pigments are typically prepared by passing hydrogen sulphide through cadmium sulphate solution. This yields the pigment as a precipitate. The final shade of the pigment is controlled by varying the precipitation conditions, and the strength and brightness of the colour are developed by calcining at high temperature. Cadmium yellow pigments approximate in composition to CdS, and the properties of the yellow pigment are similar to those of the cadmium reds previously discussed.

Calcium Plumbate, $2CaO \cdot PbO_2$

Pure calcium plumbate is a white synthetic inorganic pigment, but, in its normal commercial form, the pigment has a light buff colour. For convenience, however, it will be included in this section. Calcium plumbate is prepared by heating a mixture of calcium carbonate, $CaCO_3$, and lead monoxide, PbO, in air at 700°C, and the chemical composition of the derived calcium plumbate approximates to $2CaO \cdot PbO_2$.

Calcium plumbate is basic in nature and hence it reacts with paint media with a high acid value to form lead soaps. These soaps improve film mechanical properties, and, providing that the acidity of the binder is not excessive, the stability of the liquid paint is reasonably good. The pigment is not used to impart colour to paints but rather as a protective pigment in primers for steel and galvanised steel. In a similar fashion to other lead containing pigments used in anti-corrosive paints, its mechanism of protection is associated with polarisation of the corrosion reactions of the substrate metal. Although calcium plumbate is an effective anti-corrosion pigment, its toxic nature limits its use in coating systems.

3.2.4 Green Pigments

Chromium Oxide, Cr_2O_3

Chromium oxide, Cr_2O_3, is a synthetic inorganic pigment with a dull green colour. The pigment is prepared by heating sodium dichromate, $Na_2Cr_2O_7$ in the presence of hydrogen.

This pigment has good stability to light, heat, acids and alkalies, but paints based solely on chromium oxide are characterised by low opacity. Chromium oxide can be used in all types of paint systems where high chemical resistance and outstanding light fastness are required. The pigment also finds use as a colourant for cement and concrete.

Lead Chrome Green, $PbCrO_4:KFe(Fe(CN)_6)$

Lead chromate greens, alternatively known as Brunswick greens, are synthetic inorganic pigments varying in shade from grass green to deep green. The pigment is prepared either by dry grinding lead chrome yellow and Prussian blue (see later) or by a wet process involving the co-precipitation of lead chrome and the blue pigment. The actual shade of green is dependent upon the Prussian blue content of the pigment, with addition levels of *c*. 2% producing the lightest shade of green, and higher levels, *c*. 35%, producing the deepest shade.

Lead chrome green pigments have good opacity in films, but they tend to deepen in colour upon atmospheric exposure. Furthermore, they have poor stability to alkaline environments. It is also possible, in certain media, for the two constituent colours of the green pigment to separate out and normally the blue component 'floats' to the surface of the film. This phenomenon is known as floating or flooding.

Lead chrome greens are typically used as colouring agents in all types of paint system where the presence of lead can be tolerated.

3.2.5 Blue Pigments

Prussian Blue, $KFe(Fe(CN)_6)$

Prussian blue is a bright blue pigment, of complex composition, which is normally prepared in a two stage process. In the first stage, a solution of potassium ferrocyanide is added to a solution of ferrous sulphate, then the precipitate thus formed is separated and oxidised to the pigment Prussian blue. The pigment is referred to as potassium ferricferrocyanide and its composition approximates to $KFe(Fe(CN)_6)$.

Prussian blue is a pigment with high staining power but it has low opacity. The pigment is fast to light and it is not readily attacked by acids. However, the colour of Prussian blue is immediately discharged by contact with alkali, leaving a brown coloured oxide of iron. It also discolours in a similar fashion when heated to moderate temperature.

Prussian blue is used as a colouring pigment in many types of paint systems and is also used in the production of lead chrome greens.

Ultramarine Blue, $3Na_2O_3.3Al_2O_3.6SiO_2.2Na_2S$

Ultramarine blue is a complex synthetic inorganic pigment, although the pigment does exist in the natural form as the semi-precious mineral lapis-lazuli. However, pigments derived from the naturally occurring ultramarine are only of historic interest.

Synthetic ultramarine is a complex aluminosilicate, $3Na_2O_3.3Al_2O_3.6SiO_2.2Na_2S$, and the colour of the pigment is attributable to the presence of sulphur. The pigment is prepared by calcining an intimate mixture of sodium carbonate, china clay, sulphur and silica together with some organic resinous material such as rosin. The conversion of the reaction mixture to the final pigment form is a lengthy process taking some weeks. By varying the ratio of the raw materials, it is possible to alter the shade of the blue pigment; typically, reducing the silica content results in a green shade blue, whilst higher contents of silica produce redder shade blues.

Ultramarine blue is a coarse textured pigment prone to hard settlement when dispersed in paints. The pigment produces the reddest shades of blue available although it has poor tinting strength in paints. It has good light fastness, heat resistance and is unaffected by alkalies although it is rapidly decomposed by even dilute acids. Consequently, Ultramarine blue can be used as a colourant in paints designed for application to alkaline substrates and it is also used as a cement and concrete colourant.

3.2.6 Black Pigments

Black Iron Oxide, Fe_3O_4

Black synthetic iron oxides are produced by oxidation of ferrous hydroxide obtained from the action of alkali on ferrous sulphate solution and the composition of the pigment corresponds with that of magnetite, Fe_3O_4. The pigment is of fine texture, but has only low tinting strength in paint and allied products. The colour of the pigment is reddened by heat. Black iron oxides possess the high chemical resistance characteristics of the other iron oxide pigments.

Black iron oxide is mainly used as a colouring pigment in fillers, primers and undercoats, but it is not of sufficient quality for use in finishing systems.

3.3 METALLIC PIGMENTS

Aluminium Powder

Aluminium in powder form is obtained by grinding coarse aluminium powder in a ball-mill under an inert atmosphere. The cascading action of the balls flattens the powder into platelets of the required size. Aluminium powder is available in two forms, known as leafing and non-leafing grades, which are distinguished by their behaviour when incorporated in paints. The leafing grades of aluminium powder have the property of floating to the surface of the applied paint where they tend to orientate parallel to the plane of the film. This grade of aluminium is prepared by coating the powder with an additive, normally stearic acid, which imparts the requisite degree of surface tension to the particles. The stearic acid coating is usually applied during the ball milling operation.

Non-leafing grades of aluminium powder do not possess the required surface tension to leaf and consequently the aluminium platelets are randomly located within the paint film.

The two grades of aluminium powder are used in differing applications. The ability of leafing-grade powders to orientate themselves in a close-packed layer adjacent to the film surface provides the paint film with an efficient moisture barrier so that this grade is used in all stages of protective coating systems for metals and wood. In such applications, leafing aluminium is normally used at relatively low levels of addition, either as the sole pigment or in admixture with other pigments.

The non-leafing grade is primarily used, at a low level of addition, in the specialised decorative finishes such as the polychromatic car body finishes and finishes for instrument cases. The presence of non-leafing grade aluminium in such coatings imparts an aesthetically pleasing sparkle to the finish.

Zinc Powder

Zinc dust is prepared by distilling zinc oxide, ZnO, in the presence of carbon. This yields metallic zinc dust in a particulate form.

The main use of zinc dust is in the preparation of zinc dust primers for steelwork where they are used as the sole pigment. In this application very high levels of addition are used and in-can settlement can be a problem. High zinc contents are required to provide electrical conduction between the zinc particles (see Section 7.1.1.2). Zinc dust primers provide a high degree of corrosion protection to the steel substrate due to the tendency of the metallic zinc to corrode preferentially to steel (that is, cathodically protect the steel) in corrosive conditions.

Lead Powder

Metallic lead powder is prepared by arcing lead electrodes under water and the resultant aqueous sludge of lead powder is normally flushed into an oil medium, that is, where oil replaces the water and provides an oil-suspension.

Lead powder is mainly used in the preparation of protective primers for steelwork where it is normally present as the sole pigment. The addition level in such applications is very high and, in a similar fashion to zinc dust primers, settlement during storage of the liquid paint can be a problem. The mechanism by which metallic lead primers protect steelwork is complex, involving the formation of lead soaps within the film, which can restrict the movement of permeants. They have a barrier action and provide a degree of sacrifical protection similar to that of zinc dust.

3.4 ORGANIC PIGMENTS

3.4.1 Red Pigments

Toluidine Red

The Toluidine reds, alternatively known as Helio reds, are a class of organic compound known as insoluble azo dyes, that is, dyes insoluble in water. Toluidine pigments are a widely used class of colourant characterised by bright, clean colours, good hiding powder when used in full shade and high colour stability in the presence of light. However, when reduced with other, typically white, pigments, the colour stability of the resultant colour tends to be decreased.

As with all organic pigments, the manufacture of Toluidine reds involves somewhat sophisticated techniques and the structure of the resultant pigment molecule is itself complex. The Toluidine reds are prepared by a coupling reaction between diazotised m-nitro-p-toluidine and β-naphtol, and the pigment structure can be represented as:

The colour of the final pigment shade can be adjusted by modification of the reaction conditions and/or the reactants.

The good colour stability of the full shade pigment enables the Toluidine reds to be used as colouring pigments in decorative paints for exterior exposure. The stability of the pigment to heat up to c. 180°C is good, provided that the exposure time is restricted. Consequently, these pigments can be used in paints cured by stoving. Prolonged exposure to high temperature, however, results in severe discolouration of the naphthol component of the molecule.

Toluidine reds are soluble in aromatic solvents and are slightly soluble in alcohols and other aliphatic solvents. This solvent solubility results in the phenomenon of bleeding, whereby organic pigments in paint films can be solubilised and carried through subsequently applied paint coats by the solvents used in the paint formulation. This bleeding phenomenon results in discolouration of paint films and the effect is particularly, although not necessarily, associated with non-convertible type and stoving cured systems.

A modification of Toluidine red is typified by the pigment Permanent Red 25. This pigment has a similar composition to Toluidine red although it possesses different performance characteristics and is yellower in shade. Permanent Red 25 is prepared by coupling diazotised 2,4-dinitroaniline with β-naphthol and the structure can be represented as:

$$NO_2\text{-}\underset{}{\bigcirc}\text{-}\underset{NO_2}{}N=N\text{-}\underset{OH}{\bigcirc\bigcirc}$$

This pigment has good light fastness, even in reduction, although as with Toluidine red, Permanent Red 25 is slightly soluble in solvents. Permanent Red is not normally used in stoving finishes but is widely employed as a colouring pigment in decorative air drying finishes. In common with Toluidine red, the pigment possesses good resistance to acids and alkalies and, consequently, can be used in chemically resistant paints and in paints for application to alkaline surfaces. As a class, they are non-toxic and can be used to replace lead chrome type colouring pigments.

Arylamide Red

The Arylamide reds are a class of organic pigments varying in shade from orange through red to deep crimson. The pigments of this class are also members of the insoluble azo group and a typical Arylamide red is Permanent Red FRLL. This is prepared by coupling diazotised 2,5-dichloroaniline with β-hydroxy naphthoic acid-*o*-anisidine. The structure of the pigment can be represented by:

$$\underset{Cl}{\overset{Cl}{\bigcirc}}\text{-}N=N\text{-}\underset{}{\bigcirc\bigcirc}\overset{OH}{}\text{-}CONH\text{-}\underset{OCH_3}{\bigcirc}$$

The Arylamide reds are characterised by an extremely high degree of colour stability to light even when reduced with considerable quantities of white pigment. As a class, they are stable to temperatures in the region of $140°C$ and hence can be used as colouring pigments in stoving cured paints. Their colour stability enables this class of pigment to be used at all levels of reduction in decorative coatings for exterior exposure. They are slightly soluble in solvents, resistant to acids and alkalies and are non-toxic.

3.4.2 Yellow Pigments

Hansa Yellow

The Hansa yellow pigments are another variant of the insoluble azo group and they are produced in a range of colours, namely from orange to green shade yellows.

A typical Hansa yellow is prepared by coupling diazotised *m*-nitro *p*-toluidine and acetoacetaniline and the resultant structure can be represented by::

$$CH_3-C_6H_3(NO_2)-N=N-CH(COCH_3)(CONH-C_6H_5)$$

As a class, the Hansa yellows are characterised by good colour stability to light in full shade, although stability is lost fairly rapidly on reduction with (white) pigments. They have low opacity in paint films but, due to their non-toxic nature, they are widely used to replace lead chromate pigments in, for example, paints for children's toys. Hansa yellow pigments are soluble in aromatic hydrocarbons, ketone and ester solvents but are only slightly soluble in aliphatic hydrocarbons. Consequently, Hansa yellows are widely used as colouring pigments in air-drying decorative coatings containing aliphatic hydrocarbons and also in emulsion paints.

Benzidine Yellow

Benzidine pigments are members of the insoluble azo class of pigments and they are prepared in a range of shades from red to yellow. Benzidine yellow imparts greater opacity to paints than the Hansa range of yellows and, furthermore, they are insoluble in most solvents used in paint formulations. As a class, they are non-toxic, possess good acid and alkali resistance and are heat stable up to $c.$ $140°C$. However, they have a low colour stability on exposure to light, even in full shade. This consequently limits their use in exterior coatings although, because of their solvent resistance, they are widely used as colouring pigments in stoving cured paints for interior use. They are also used as replacements for lead chrome pigments.

A typical Benzidine yellow is prepared by coupling diazotised 3,3-dichlorobenzidine with acetoacetanilide and the structural formula can be represented by:

$$\text{CH}_3-\text{CO}-\text{CH}(\text{CONH-C}_6\text{H}_5)-\text{N}=\text{N}-\text{C}_6\text{H}_3(\text{Cl})-\text{C}_6\text{H}_3(\text{Cl})-\text{N}=\text{N}-\text{CH}(\text{CONH-C}_6\text{H}_5)-\text{CO}-\text{CH}_3$$

3.4.3 Green Pigments

Pigment Green B

Pigment Green B is an organic pigment of the class known as toners. Toners are a sub-division of organic pigments and, unlike the pigment or azo dyestuffs described above, they are not purely organic in nature. Toners can be defined as being salt-like compounds derived either by the reaction between a metal or the salt of a metal and a dyestuff containing an acid group, or through the reaction of a basic dyestuff with a complex acid. As a class, toners are more resistant to bleeding in solvents than the purely organic dyestuffs.

Pigment Green B is prepared by the reaction of α-nitroso-β-naphthol with ferrous sulphate and the structure can be represented by:

$$\left[\text{naphthoquinone-N-O} \right]_2 \text{Fe}$$

Pigment Green B is a blue shade green imparting good opacity to paints. The resistance of the pigment to both solvents and alkalies is very good, but it has poor acid resistance. The pigment is widely used as a colouring pigment in decorative finishes, particularly where resistance to alkali is a requirement, and, due to its heat stability, it can be used in stoving cured paint finishes. It is also be used as a cement and concrete colourant.

3.4.4 Blue Pigments

Phthalocyanine Blue

The Phthalocyanine pigments are a class of organic pigment varying in colour from a red shade blue to a yellow shade green. Phthalocyanine pigments are widely used in paints. They are characterised by high tinting strength and opacity together with excellent colour stability on exposure to light. As a class, the pigments are non-toxic, heat stable up to temperatures of c. $500^{\circ}C$ and they are resistant to most chemicals although they are sensitive to strong acids. These pigments are also insoluble in most solvents used in paint and hence are not prone to bleeding.

A typical Phthalocyanine blue pigment is prepared by heating together an intimate mixture of phthalic anhydride, urea and cupric chloride at a temperature of c. $220^{\circ}C$ for 5 hours, and the structure can be represented by :

The green shade Phthalocyanine pigments are prepared by chlorinating the benzene rings, and the pigments can be rendered water soluble for use in emulsion paints by sulphonating the benzene rings.

The excellent stability of the phthalocyanine pigments permits their use as colourants in all forms of decorative and industrial coating systems.

3.4.5 Black Pigments

Carbon Black

Carbon blacks are organic pigments produced by the incomplete combustion, or carbonisation, of mineral, vegetable and animal matter. With certain types of carbon black, there is an inorganic residue in the pigment which adulterates the properties of the pigment. For example, carbon black pigment prepared from the mineral hydrocarbons such as oil and natural gas have a relatively high carbon content and such grades are characterised by having a very fine texture, pure colour and high opacity in paints. Black pigments with lower carbon contents are prepared from vegetable oils or coal-tar distillates and these have an inferior colour and opacity compared with the high carbon blacks.

As a class, carbon black pigments are insoluble in solvents, stable to acids and alkalies and are fast to light even at high reductions. Carbon blacks are of high tinting strength and they provide high opacity in paints at the low levels of addition that are normally used. They are used as colouring pigments in all types of decorative and industrial paints, and the high carbon grades are suitable for use in finishing systems.

3.5   EXTENDERS

Barytes, $BaSO_4$

The extender barytes is a naturally occurring inorganic mineral composed of barium sulphate, $BaSO_4$. The mineral is mined, ground and purified prior to its use in paints and it is characterised by a very high resistance to acid and alkalies. However, barytes is very hard and, consequently, is difficult to disperse in paints by the conventional grinding techniques and this, together with its high specific gravity, can lead to in-can settlement. In a similar fashion to other extender materials, barytes is transparent in oil and other resinous media and, consequently, it can be used in paints based on solvent soluble binders without substantial effect on either their colour or opacity.

Barytes can be used in fillers and stoppers for all surfaces and in decorative and industrial primers and undercoats, where it acts both as a chemically inert filler and as a means of increasing the mechanical strength of the film. In common with all extenders, however, an excess of barytes in paint formulations can be detrimental to the subsequent durability of the film and consequently, barytes (and certain other extenders) is only rarely used in exterior finishing coats.

Barium sulphate can also be prepared synthetically as a product of the reaction between barium sulphide and sodium sulphate solutions. This synthetic form of barium sulphate is known as blanc fixe and it has similar characteristics and applications to those of natural barytes.

Whiting, $CaCO_3$

Whiting, or chalk, is a mineral extender composed of calcium carbonate, $CaCO_3$, and it can be obtained either naturally, as limestone, or synthetically. A typical preparatory route involves passing carbon dioxide through calcium hydroxide solution, with the calcium carbonate being obtained as a precipitate. The synthetic form of calcium carbonate is sometimes called Paris White.

Whiting (and Paris White) can be used as a filler in a wide range of coating systems but, due to its sensitivity to acid environments, it is not normally incorporated in paint systems designed for exposure in severe conditions. This extender is widely used in interior and exterior quality emulsion and water paints, where it imparts some pigmentary properties. In solvent containing paints, whiting is transparent, and small additions can be beneficial in reducing settlement of paints during storage. Whiting is also widely used in undercoats as well as in interior quality flat finishes.

China Clay, $Al_2O_3.2SiO_2.2H_2O$

China clay or kaolin is a hydrated silicate of aluminium approximating in composition to $Al_2O_3.2SiO_2.2H_2O$. The extender is obtained naturally from the mineral granite by means of high pressure water jets which remove a sludge of kaolin from the face of the mineral deposit. The extender, china clay, is obtained as a refined product of very fine particle size after a levigation process.

China clay is used as a filler in solvent containing primers and undercoats, but addition levels are not normally high, as the fine particle size of the material can adversely affect the flow properties of paints. This extender also exerts a flatting effect on the gloss level of paints and, consequently, it is used as a flatting agent in undercoats and eggshell sheen finishing paints. The fine particle size of china clay, however, assists in reducing settlement of paint during storage. Even at relatively low levels of addition, china clay may reduce the strength of films and, consequently, it is only of limited applicability in coating systems for exterior use.

Mica, $K_2O \cdot 2Al_2O_3 \cdot 6SiO_2 \cdot 2H_2O$

Mica is a complex potassium aluminosilicate with the approximate composition of $K_2O \cdot 2Al_2O_3 \cdot 6SiO_2 \cdot 2H_2O$, and the extender is used in the form of thin platelets obtained from disintegration of the crude ore Muscovite. Care has to be exercised during production to avoid crushing of the platelets.

In paints, the plate-like or lamellar structure of mica results in a leafing effect analogous to that of aluminium powder and, consequently, the extender is used to reduce the rate of water penetration through paint films. Mica is one of the few extenders used in exterior finishing paints, not only because it reduces water permeability, but also because paint films containing mica are less prone to fail by cracking and checking, and have improved weather resistance. Mica can be incorporated in all types of coating, but it is generally only used in protective paints since the lamellar structure of the extender can reduce the aesthetic appearance of decorative finishes.

Talc, $H_2Mg_3(SiO_3)_4$

Talc or French Chalk is a hydrated magnesium silicate, approximating in composition to $H_2Mg_3(SiO_3)_4$, and the extender is obtained from the crude mineral by a relatively simple grinding process.

Talc occurs in a lamellar form similar to that of mica but mixed with fibrous particles. This fibrous structure provides film reinforcement and aids flexibility, whilst the presence of platelets results in improved film water resistance. Consequently, this extender finds similar applications to those of mica, namely in protective paint systems for all substrates where high durability is a primary requirement.

3.6 BIBLIOGRAPHY

Paint Technology Manual, Part 6: Pigments, Dyestuffs and Lakes, A.R.H. Tawn (ed.), Chapman and Hall, London (1966).

Pigment Handbook, Volumes I, II & III, T.C. Patton, J. Wiley, New York (1973).

4 Additives and solvents for paints

4.1 PLASTICISERS

The main function of a plasticiser is to increase, and maintain, film flexibility, particularly in paints based on binders which, in the absence of plasticisation, tend to be brittle. Plasticisers can either be added physically to the polymer, generally during paint manufacture, or they can be chemically incorporated into the polymer molecule by copolymerisation techniques. The former mode of plasticisation is known as external plasticisation and the latter as internal plasticisation. Examples of internal plasticisation have already been mentioned in Chapters 1 and 2, e.g. the copolymerisation of the flexible vinyl acetate with the brittle vinyl chloride. Consequently, discussion here will be limited to the external plasticisers.

In many coating systems such as those based on chlorinated rubbers and nitrocellulose, the plasticiser component may constitute a major proportion of the total binder content. Clearly in such instances, the nature of the plasticiser can exert a considerable influence on the performance of the paint film. Other polymers and coating systems do not require such large plasticiser additions and many polymers are inherently sufficiently flexible not to require external plasticisation.

Plasticisers are normally low molecular weight, non-volatile liquids which ideally are fully compatible with the polymer component of the film. In most applications, they should also exert a minimum softening effect on the film whilst imparting maximum effect on its flexibility. In certain instances, however, high molecular weight resinous materials are used to impart flexibility to relatively brittle systems, as for example, in the addition of alkyd resins to chlorinated rubber.

Other requirements are also of importance. In particular, plasticisers should not alter the colour of the film nor should they themselves change colour significantly on weathering. Non-toxicity is also a primary requirement in the majority of applications.

The effect of plasticisers on paint film properties can best be understood by considering their influence on the glass transition temperature, T_g, of the film. The T_g is the temperature at which the polymer component of the paint film, on cooling from elevated temperatures, changes from a rubbery melt to a glass-like solid at lower temperatures. If maintained at temperatures below their T_g, therefore, polymers tend to be rigid and brittle and, since pigmentation has little effect on the T_g, then the paint film based on the polymer will also be brittle. The brittle characteristics of polymers at temperatures below their T_g are attributable to the high strength of the intermolecular cohesive forces. These forces effectively inhibit intramolecular movement and confer rigidity upon the molecular networks.

As the ambient temperature approaches and exceeds the T_g, the molecule chains become more mobile and the film will exhibit increasing flexibility. Clearly, therefore, it is desirable for paint films to be above their T_g during service and thus, ideally, the polymeric binder should have a low T_g. However, even at normal room temperatures many polymers are below their T_g and hence would, in their unmodified state, exhibit brittle characteristics. The addition of plasticisers to such films results in the molecular chains of the polymers becoming more separated and therefore the intermolecular forces within the polymer network are weakened. This results in a reduction of the film T_g. With films based on polymers possessing a T_g substantially below room temperature, external plasticisation is not normally necessary. The glass transition temperatures of some typical polymers and polymer-plasticiser combinations are shown in Table 4.1.

Plasticisers used in paints can be broadly classified into two classes, namely primary (or solvent) plasticisers and secondary plasticisers, depending on their function. The primary plasticisers can be considered to be solvents for the polymer and, as such, the two are naturally compatible in all proportions. This type of plasticiser contains chemical groups which can interact

Resin	T_g, °C
Poly (vinyl chloride)	80
Poly (vinyl acetate)	30
PVC : PVA (95:5)	50
Nitrocellulose	50
Nitrocellulose + Tricresyl phosphate	
80 : 20	-10
60 : 40	-30
Non-oil-modified alkyd (Glyptal phthalate resins)	85
60% Linseed oil alkyd	12
Oil-soluble phenolic	80
Linseed oil phenolic varnish	
1 : 1	30
3 : 1	10
Linseed oil	-18

TABLE 4.1 Glass Transition Temperatures of Resin Systems

with the polymer, such that the relatively low molecular weight plasticiser may be introduced into and between the high molecular weight chains of the polymer to form a less rigid structure. Such plasticisers however, normally reduce the strength characteristics of the paint film.

The class of plasticisers known as secondary plasticisers are not solvents for the polymer and are often only compatible over a limited range of addition levels. These plasticisers are non-reactive with the polymer and they act only as lubricants, that is, physically separating the molecular chains within the polymer so that they can slide past each other when film flexing occurs. This class of plasticiser has a lower adverse effect on film strength than the primary plasticisers. Under certain conditions, however, they can migrate or be leached from between the polymer chains, leading to premature loss of film flexibility.

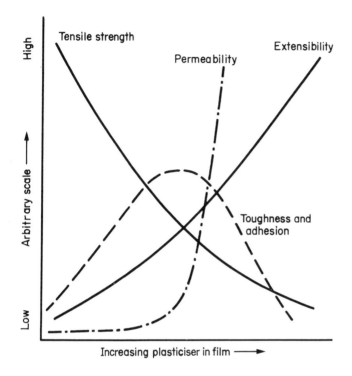

Figure 4.1   Effect of plasticisation on paint film characteristics

The effect of plasticisers on the most important characteristics of paint films is shown schematically in Figure 4.1. Due to the wide variety of polymers requiring plasticisation and the many classes of materials that can be used to impart plasticity, this diagram is necessarily a simplification. In general therefore, film toughness and adhesion reach a peak at a certain level of plasticisation and then decrease. Film permeability characteristics undergo a marked increase as a certain level of plasticisation is exceeded. Increasing amounts of plasticiser produce increasing film extensibility and reduced film tensile strengths.

The properties and uses of several important types of plasticiser are summarised below.

Dibutyl Phthalate

Dibutyl phthalate, DBP, is a widely used plasticiser, characterised by good compatibility with a wide range of resins and moderate resistance to yellowing on exposure to light. It is, however, slightly volatile and this tends to limit its usefulness. DBP is one of the most important plasticisers for nitrocellulose resins, where it is used in various proportions depending upon the service application of the paint composition. Generally, however, addition levels of 20-50% of the weight of nitrocellulose are typical.

A further important use of DBP is in conjunction with polyvinyl acetate homopolymer emulsions. Addition levels of 10-20% are typical and the resultant emulsion is widely used as a binder in emulsion paints and as a general purpose adhesive in the building industry.

Dioctyl Phthalate

Dioctyl Phthalate, DOP, is less volatile than DBP and has good stability to light and heat. The compound is widely used in nitrocellulose finishing systems and also in polyvinyl chloride organosol and plastisol systems.

Triphenyl Phosphate and Tricresyl Phosphate

These two compounds are mainly used in nitrocellulose finishing systems where, apart from their plasticising characteristics, they tend to reduce the flammability of the resultant films. Both types of plasticiser have minimal effect on the strength of nitrocellulose films. Tricresyl phosphate, however, yellows on exposure to light and consequently it is not used where film colour stability is a prime requirement.

Trichloroethyl Phosphate

Trichloroethyl phosphate, TCEP, is another widely used organophosphate which also possesses the property of reducing the flammability of nitrocellulose based paints. Furthermore, TCEP shows little tendency to yellow on exposure to light and, in contrast to other phosphate plasticisers, it is effective at low temperatures.

Butyl Stearate

Unlike the plasticisers discussed above, butyl stearate is a secondary plasticiser for nitrocellulose and hence, when used in nitrocellulose coatings, it maintains and improves the scratch and abrasion resistance of the film. Since this plasticiser upgrades film hardness, it is used in nitrocellulose formulations where the final gloss of the film is developed by a post-application polishing operation. Butyl stearate is also effective at low temperatures.

Chlorinated Paraffins

The chlorinated paraffins are a class of plasticiser used primarily in conjunction with chlorinated rubber resins. This class of plasticiser is chemically inert and, when used in combination with chlorinated rubber, it does not detract from the high chemical resistance characteristics of the resin. Addition levels of up to 50% of the resin content are typical.

4.2 SOLVENTS

Solvents are volatile liquids added to paints and allied products in order to dissolve the resin component and to modify the viscosity of the coating. To be effective, a solvent must fulfil certain criteria. It must yield a solution of suitable viscosity to suit the storage and application requirements of the liquid coating. It should have the correct evaporation rate and it must deposit a film with optimum characteristics. It should also have an acceptable odour, minimal toxicity and reasonable cost.

With paints based on convertible resins, solvents are primarily added to enable the coating to be applied by the appropriate technique. Solvents in coating compositions based on non-convertible resins perform a more complex function since not only do they control the application characteristics of the coating but they also perform an important role in determining the drying time and quality of the resultant film. In such instances, it is common practice to use a blend of two or more solvents to achieve the requisite performance characteristics from the coating. Typically, both a true solvent for the particular resin and a diluent, that is, a liquid that by itself is not a true solvent for the resin but which aids dissolution when used in conjunction with the true solvent, are used. Diluents are also cheaper than the true solvents required for many types of non-convertible media.

Two important characteristics of solvents are their solvent power or ability to dissolve specific resins, and their evaporation rate, that is, the relative speed with which they leave the applied coating after application.

4.2.1 Solvent Power

The molecular structure of the resins used in the paint industry typically consist of long chains with pendant side groups, and all of these chains within the resin mass are in a constant state of movement due to Brownian motion. When movement brings two chains into close proximity, short range attractive forces, known as secondary valence bonds, can become established, resulting in restricted chain movement and the formation of a viscous system. The solution of a resin in a solvent can be considered as a form of 'invasion' by the solvent molecules as they fill spaces made available by chain movements. The solvent molecules thereby increase the distance between the molecular chains of the resin and thus restrict their ability to form short range valence bonds. Increasing additions of solvents cause increasing chain separation and result in reduced interaction and hence lower solution viscosity.

In practice, it is found that resins will only dissolve in specific solvents, and traditionally selection of solvents was performed on an empirical 'like dissolves like' basis. However, an exact method has been developed for selecting a solvent or combination of solvents for a particular resin and this is known as the solubility parameter concept. This concept is based on the principle that the free energy, (ΔG), becomes negative when a polymeric substance is dissolved by a solvent. This concept of free energy is dependent upon the difference between the heat of mixing, (ΔH), and the product of the temperature and the degree of disorder or entropy, (ΔS), that is:

$$\Delta G = \Delta H - T\Delta S \qquad (4.1)$$

This is known as the Gibbs free energy equation and holds for all chemical processes.

When a polymer dissolves in a solvent there is a large increase in the disorder of the system and the $T\Delta S$ product in Equation 4.1 has a relatively large value. If the magnitude of ΔH is less than that of $T\Delta S$, then the dissolution process will occur spontaneously, and the term ΔH will approach zero when the intermolecular forces between the solvent and the resin molecules are similar. These intermolecular forces are expressed as cohesive energy density and the square root of this expression is the solubility parameter, δ. Solubility parameters have been determined for all of the solvents and resins commonly used in paint formulations and some examples are shown in Tables 4.2 and 4.3. Normally, single numbers are quoted for solvents, whereas a range of values are given for resins. If the δ value of a solvent lies within the range for a resin, then dissolution can occur. The solubility parameters of solvent combinations are normally taken as the average value of the parameters of the components based on the percentage composition by volume.

Solvent	Hydrogen bonding		
	δ_p	δ_m	δ_s
Aliphatic Hydrocarbons:			
White spirit	7.5		
Aromatic Hydrocarbons:			
Toluene	8.9		
Xylene	8.8		
Esters:			
Ethyl acetate		9.1	
Butyl acetate		8.5	
Ketones:			
Acetone		10.0	
Methyl isobutyl ketone		8.4	
Glycol Ethers:			
Ethylene glycol monoethyl ether		9.6	
Alcohols:			
Ethyl alcohol			12.7
Butyl alcohol			11.4

TABLE 4.2 Solubility Parameters of Some Common Solvents

	Solubility parameter ranges in solvents with hydrogen bonding		
	δ_p	δ_m	δ_s
Alkyds			
68% oil length linseed	7.0-11.1	7.4-10.0	9.5-11.4
52% oil length linseed	7.0-11.1	7.4-11.9	9.5-11.4
44% oil length linseed	8.1-11.1	7.4-11.9	9.5-11.4
Vinyls			
Vinyl chloride : vinyl acetate	9.3-11.1	7.8-13.5	0
Vinyl chloride : vinyl acetate: dibasic acid	10.6-11.1	7.8-12.2	0
Polyvinyl butyral	0	8.9-10.8	9.5-14.5
Polyvinyl chloride	10.6-11.1	9.3-9.9	0
Epoxies			
Bisphenol A modified solid epoxies:			
Epoxy equivalent 450-500	10.6-11.1	8.9-13.3	0
810-900	0	8.4-13.3	0
1700-2000	0	8.4-13.3	0
2500-4000	0	8.4-9.9	0
Miscellaneous			
Drying oil epoxy ester	8.5-11.1	7.8-9.9	0
Chlorinated rubber	8.5-10.6	7.8-10.8	0
Polymethyl methacrylate (Acrylic)	8.9-12.7	8.5-13.3	0
Cellulose nitrate	11.1-12.7	7.8-14.7	12.7-14.5
Oil free polyester	8.6-11.1	7.4-11.9	9.5-11.4
Polyurethane, TDI - polyester adduct	8.8-11.1	8.7-11.9	0

TABLE 4.3 Typical Solubility Parameter Ranges of Various Classes of Resins

However, one other factor is required to predict solubility accurately and that is the degree of hydrogen bonding of the solvent. Three classes are normally recognised, namely solvents that are strongly hydrogen bonded, δ_s, (e.g., alcohols), solvents that are moderately hydrogen bonded, δ_m, (e.g., esters and ketones) and solvents that are poorly hydrogen bonded δ_p (e.g., aliphatic and aromatic hydrocarbons).

An indication of the use of solubility parameters can be gained from the following example. Chlorinated rubber has the solubility parameters $\delta_s = 0$, $\delta_m = 7.8 - 10.8$ and $\delta_p = 8.5 - 10.6$. Reference to Table 4.2 indicates that dissolution will be effected by xylene or toluene, which have δ_p values of 8.8 and 8.9 respectively, and by all of the δ_m solvents, esters, ketones and glycol ethers shown in the table. However, the $\delta_s = 0$ value of chlorinated rubber indicates that it will not be soluble in any alcohol solvent.

In practice, it is likely that a combination of solvents will be used to effect solution of such a resin, both from considerations of cost and to optimise paint viscosity and drying time. This can readily be achieved providing that the solubility parameters and the degree of hydrogen bonding of the solvents lie within the range of the resin.

The amount of diluent (non-solvent) that a resin will tolerate can be estimated by calculating the average δ value of the solvent-diluent mixture that will lie within the mid 80% of the δ range of the polymer.

If solvent resistance is a requirement of a coating system, then the resin component of the film should be selected to have a δ value as far removed as possible from the δ value of the solvent to be resisted. Conversely, compatibility of two or more resins is normally assured if the midpoints of the ranges of their respective δ values differ by no more than 1 unit.

4.2.2 Evaporation Rate

The rate at which solvents evaporate from the applied coating is of importance, particularly in formulations based on non-convertible resins such as chlorinated rubber and vinyl chloride-acetate copolymers. Solvents with too rapid an evaporation rate from the applied liquid coating can cause such problems as poor flow and inferior film integration, with the latter often leading to poor coating durability. Conversely, solvents which evaporate too slowly can result in excessively long drying times, which may hinder production line finishing operations, or may result in excessive flow and sagging of the applied coating. In practice, too fast an evaporation rate is the greater problem, since paint formulations can usually be modified to offset excessive flow characteristics, for example, by introducing a degree of thixotropy into the system (see Section 4.6).

A basic principle is that coatings applied by brush are formulated with solvents possessing relatively slow evaporation rates, whereas coatings designed for spray application require more rapidly evaporating solvents.

The rates of evaporation of solvents are often expressed as numerical values relative to some other solvent, the evaporation rate of which is taken as a standard, Table 4.4. Evaporation rates are determined on samples of the pure solvent, but in practice it is found that the presence of solids, and in particular resins, can reduce these values, that is, slow down the rate of evaporation. This effect generally seems to be most pronounced for the aliphatic and aromatic hydrocarbon solvents. The characteristics and uses of the more important solvents commonly used in paint manufacture are outlined below.

Solvent	*Evaporation rate by volume, where butyl acetate* = 100
Acetone	944
Ethyl acetate	480
Ethyl alcohol	253
Toluene	214
Methyl isobutyl ketone	164
Butyl acetate	100
Xylene	73
Butyl alcohol	36
Ethylene glycol monoethyl ether	30
White spirit	*c.* 18

TABLE 4.4 Evaporation Rates of Some Common Solvents

4.2.3 Hydrocarbon Solvents

White Spirit

White Spirit has a variable composition and the grades commonly used in paints are substantially aliphatic in nature but have an aromatic content of approximately 15-18%. The solvent is characterised by a slow evaporation rate together with a mild odour and it will dissolve most oils, natural resins, oleoresinous varnishes and alkyd resins with an oil length greater than 50%. As a result, white spirit is widely used as the (sole) solvent in many types of decorative and protective coatings based on these oxidative media where application is by brush.

White Spirit is also a commonly used general purpose cleaning and degreasing solvent and it is the most widely used of all solvents in paint formulations.

Toluene, $C_6H_5.CH_3$

Toluene is commonly used, in conjunction with other solvents, in the formulation of air drying vinyl copolymer and chlorinated rubber coatings. Toluene is also used in the preparation of nitrocellulose based coatings where it acts as a diluent.

Xylene, $C_6H_4.(CH_3)_2$

Xylene is widely used in industrial coating systems, being a solvent for polyurethanes, chlorinated rubber, vinyl copolymers, alkyds and modified alkyds with oil lengths less than 50%. This is largely due to its good solvent power and moderate evaporation rate which is slow enough to facilitate good flow but sufficiently fast to permit application, by spray, of relatively high film thicknesses without sagging. Xylene is particularly suited for use in stoving-cured coatings since the evaporation rate permits a satisfactory 'flash off', that is the loss of the bulk of the solvent from the film prior to the coated articles being introduced into the oven.

4.2.4 Esters and Ketones

Ethyl Acetate, $CH_3COO.C_2H_5$

Ethyl acetate is characterised by a rapid evaporation rate and a mild, rather pleasant fruity odour. Ethyl acetate is one of the best solvents for nitrocellulose and it is used primarily in this application. Nitrocellulose coatings containing ethyl acetate have a higher tolerance for aliphatic hydrocarbon diluents than coatings containing ketone type solvents. Furthermore, the mild odour of ethyl acetate enables it to be used in preference to ketones, especially in nitrocellulose coatings designed for the retail market.

Butyl Acetate, $CH_3COO.C_4H_9$

Butyl acetate has a moderate evaporation rate and a pronounced fruity odour. It is one of the most important solvents for nitrocellulose and its modifying resins and this constitutes its major use in the paint industry.

Acetone, $CH_3.CO.CH_3$

Acetone, dimethyl ketone, is a powerful solvent with an extremely fast evaporation rate. This solvent is commonly used in vinyl copolymer and nitrocellulose formulations, although it is normally added in relatively small amounts (to a blend of other solvents) where its high solvent power and fast evaporation rate might usefully modify the application and film formation properties of the coatings.

Methyl Isobutyl Ketone, $CH_3.CO.CH_2CH(CH_3)_2$

Methyl isobutyl ketone (MIBK) is characterised by high solvent power and a moderate evaporation rate. MIBK is widely used as a solvent for epoxy, polyurethane and nitrocellulose coating systems, although it is generally modified with slower evaporating solvents to facilitate the flow and film formation characteristics of the binder.

4.2.5 Glycol Ethers and Alcohols

Ethylene Glycol Monoethyl Ether, $C_2H_5.O.CH_2CH_2OH$

Ethylene glycol monoethyl ether, more commonly known as 'Cellosolve' is a solvent for many resin systems and, due to its slow evaporation rate, it is frequently added to coating formulations designed for brush application that will not tolerate additions of aliphatic hydrocarbons.

Ethyl Alcohol, C_2H_5OH

The pure solvent ethyl alcohol, adulterated with additions of toxins, is commercially available as industrial alcohol (or methylated spirits). This solvent is used in the preparation of nitrocellulose coatings, in etch primers based on poly (vinyl butyral), and in furniture lacquers.

Butyl Alcohol, C_4H_9OH

Butyl alcohol is characterised by a pungent odour and is a solvent for many types of oils and resins. It is a solvent for alkyd and amino modified alkyds and also finds use in nitrocellulose and acrylic based coatings.

4.3 DRYING ACCELERATORS

Drying accelerators or driers are compounds added to paints based on oxidation curing resins, namely oils, varnishes, alkyds and modified alkyds, to increase the drying or curing rate of the applied film. These driers are generally oil soluble organic acid salts of various metals, the most widely used being the naphthenates of lead, cobalt and manganese. A fairly wide variety of metal salts of organic acids are, however, used as driers, less commonly used examples being zirconium and aluminium salts.

Driers function by accelerating the oxidation rate of the unsaturated oil component of the binder, facilitating cross-linking at the double bonds, although the precise reaction mechanism is complex and not fully understood. Driers are only effective in oils and resin systems containing a high degree of unsaturation, that is, drying and semi-drying oils. The curing rates of binders containing the saturated non-drying oils are not influenced by the addition of driers but are determined by cross-linking of the resin component with other reactive resins such as the amino resins.

The most active drier metals are lead and cobalt. They function in different ways within paint films and consequently combinations of these two driers are normally used in order to achieve optimum results. Driers containing cobalt salts function as surface driers, that is, they increase the oxidation rate of the outermost layer of the paint film. If a cobalt salt was used as the sole drier in a paint, then there would be a definite risk of film rivelling due to an uneven drying rate. Lead-containing driers are through-film driers, that is, they function as polymerisation catalysts within the film. Combinations of these two drier metals are commonly used in paints in order to achieve a uniform drying rate, with a ratio of

10:1, lead to cobalt salts, being typical. The use of lead driers in conjunction with cobalt driers also reduces the tendency of alkyd gloss paints to 'bloom' when drying under cold, damp conditions. Additions of oil soluble organic acid salts of calcium are also beneficial in reducing 'bloom' tendencies, although they have no effect on the drying rate of paints.

The acid radical component of the drier metal is not directly involved in the drying reaction but does have an influence on the properties of the resultant paint. Metal salts based on naphthenic acids, known as naphthenate driers, are the most widely used type of drier and they have no deleterious effects on the durability of alkyd films. Metal salts based on octoic acid also have a wide usage. Driers prepared from rosin or linseed oil fatty acids (the rosinate and linoleate driers) are not widely used in modern coatings since in contrast to the naphthenate driers, they exhibit a progressive loss of drying efficiency on prolonged storage of the bulk liquid paint. Rosinate driers also adversely affect film durability by accelerating embrittlement of the resin component, this effect being attributable to the low durability of rosin.

The drier content of paint formulations is critical and typical addition levels, calculated on the solid content of the binder, are 0.25-0.5% lead metal and 0.025% - 0.05% cobalt metal.

In recent years, there has been a growing concern over the toxicity hazard associated with lead compounds and, consequently, interest has developed in lead-free driers. Organic acid salts of zirconium are proving to be suitable alternatives to lead based driers whilst possessing the additional advantage of maintaining their effectiveness at low temperatures.

4.4 BIOCIDES, FUNGICIDES AND ANTI-FOULING ADDITIVES

4.4.1 Biocides

Any paint that contains water as a major ingredient, such as an emulsion paint, is potentially subject to attack by micro-organisms such as fungi and, in particular, bacteria. This attack leads to 'in can' spoilage of the liquid paint, resulting in such phenomena as loss of viscosity and pressure build up within the can due to gassing and putrefaction. Materials that are added to paints to obviate bacterial attack are known as biocides.

The component of emulsion paints most susceptible to bacterial and fungal attack is the thickener, typically a cellulose ether, and bacteria effect breakdown of this material by secreting extra-cellular enzymes. These enzymes are proteinaceous substances which catalyse the decomposition of high molecular weight compounds into a form which is readily degraded and assimilated by the bacteria. The biocides used in paints are generally only toxic towards bacteria and have little effect on their enzymes. Care must therefore be exercised during the manufacture of emulsion paints to ensure that the raw materials added, in particular the water, are not contaminated by enzymes since, in the presence of the latter, paint degradation can still occur even when a biocide has been included in the paint formulation.

The micro-organisms most frequently found in emulsion paints are Pseudomonas, Aerobacter, Flavobacterium and Bacillus species and such organisms are all capable of causing paint degradation.

Biocides function by affecting the basic metabolism of the bacterium and thereby interfere with enzyme synthesis. In order to do this, the biocide must be absorbed by the bacteria from the surrounding environment, that is, the aqueous phase, and thus an important requirement of biocides is that they are water soluble. Furthermore, they should remain in the aqueous phase on storage, that is, they should not be adsorbed by any pigments or extenders present in the paint.

Many materials have biocidal action in emulsion paints. Complex compounds of phenol, formaldehyde and, less commonly, mercury are used. Addition levels vary with the type of biocide but typically are 0.05% - 0.3% of the weight of paint.

4.4.2 Fungicides

Under certain conditions, paint films can become contaminated, either from the atmosphere or from the substrate, by a variety of fungal species. These can contribute to degradation and failure of the coating although, generally, they only result in an aesthetically unattractive discolouration of the surface. In certain situations, however, for example, kitchens, food processing plants, dairies and breweries, the presence of fungi on paint films constitutes a health risk even though no film damage may occur. To offset fungal colonisation and growth on paint films, toxins known as fungicides are added at the formulation stage.

Fungal growth can occur in both interior and exterior environments, the prime requirement for growth being high relative humidity and warmth. In general, paints for exterior service in temperate or cold climates are not subject to fungal attack. However, paints for use in humid, tropical climates generally require the addition of fungicides.

In contrast to the predominantly simple single celled bacteria, fungi are complex organisms and they obtain food by absorption through filamentous hyphae. Fungi require an organic source of carbon and nitrogen, together with trace amounts of other elements, oxygen and a relative humidity in excess of 70% in order to proliferate.

On an exposed paint film, these requirements for fungal proliferation are generally satisfied from the film itself, particularly from the resin component, although food may also be obtained by extraction from the underlying substrate surface or through deposition of organic atmospheric debris onto the paint film. A large variety of fungal species have been isolated from paint films, the most common being the Pullularia, Cladosporium, Phoma, Aspergillus and Penicilium species. The characteristics and applications of certain types of fungicides are summarised below.

Zinc Oxide

This is a traditional fungicide requiring high dosage levels, normally in the region of 30-40%, to be effective. Because of its pigmentary properties it is often used as combined pigment and fungicide in suitable media. The relatively poor exterior durability of zinc oxide containing paints and its somewhat variable fungicidal action, however, militates against its use in modern surface coatings.

Barium Metaborate

Good fungicidal action with barium metaborate is only achieved at addition levels of 15-20% of the total paint solids. Barium metaborate has a slight water solubility and, for exterior use, it has to be used in paint media characterised by relatively low permeability.

Organomercurials

Many organomercury compounds have found use as fungicides in paints, for example, phenyl mercury acetate. They are also effective biocides. Mercury compounds, however, have certain performance limitations in that they are leachable, are decomposed by ultra-violet light and are susceptible to blackening in sulphide containing atmospheres. In addition to these limitations, organomercury compounds are very toxic towards higher organisms and this restricts their use in current practice.

Organotin Compounds

Less toxic to higher organisms than the organomercurials, organotin compounds such as tributyl tin oxide are used as fungicides in paints, particularly those designed for interior use. Addition levels are dependent on the particular resin medium but levels of 1% on total solids are typical.

Dithiocarbamates

Dithiocarbamates have a broad spectrum of effectiveness against many types of fungi which, with their low water solubilities, leads to their wide use in both interior and exterior paints. Typical addition levels are 3-6% of total solids. In general, dithiocarbamates are not used in paints based on oxidising media since they tend to excessively increase the drying times of such binders.

Dichlorfluamide

Dichlorfluamide is a broad spectrum fungicide used in interior and exterior coating systems, addition levels in the region of 1.5-2.0% of total solids being typical. This fungicide, however, is not stable in emulsion paints, being subject to hydrolytic decomposition in the slightly alkaline conditions typical of such formulations.

4.4.3 Anti-fouling Agents

Surfaces submerged in sea water readily 'foul' with a variety of marine organisms, such as, weed, barnacles and tube worms. The degree of fouling in any one area depends on a number of factors, such as, the season of the year, the temperature of the water and the amount of light reaching the surface being fouled, as well as on the depth of immersion and the water flow rate relative to the surface. With ships, heavy underwater fouling can result in considerable economic loss through the increased frictional drag and the consequent increase in fuel consumption. Furthermore, the need for periodic dry docking in order to remove the marine growth is expensive and results in further loss of revenue during dry docking. Immersed metallic structures, when subjected to fouling, usually exhibit accelerated corrosion rates due to the establishment of differential aeration cells. Furthermore, sulphate reducing bacteria, which may be trapped within marine deposits, may also lead to enhanced corrosion of the metal.

To prevent the fouling of underwater surfaces, paints known as anti-fouling compositions are used to provide a semi-permanent surface treatment. These compositions contain soluble poisons that are toxic to a wide variety of marine fouling organisms. In use, the anti-fouling composition is applied as a sub-layer or 'undercoat' beneath other components of the coating system, and toxin is leached progressively to the surface of the paint film by dissolution in the permeating aqueous phase. The rate at which the toxin is released to the surface of the film clearly influences the useful service life of the anti-fouling composition, with too rapid a leach rate resulting in only short-term effectiveness. It is a necessary requirement, however, for the anti-fouling agent to have a certain minimum leaching rate such that at any given moment, sufficient toxin reaches the surface of the coating system to stifle attachment and growth of fouling organisms.

Compounds such as metallic copper, copper suboxide, tributyl tin oxide and mercuric oxide are commonly used as anti-fouling toxins. The addition levels of these substance are normally very high in order to provide a sufficient reservoir of poison within the paint film.

4.5 PIGMENT DISPERSING AGENTS

During paint manufacturing processes, it is necessary to achieve a uniform dispersion of solid particulate pigments and extenders within a liquid medium. This dispersion process involves breaking down of pigment agglomerates and uniformly wetting the individual' pigment particles with the paint medium. In order to wet a pigment surface, occluded gases and water present on the surface of most pigment particles have to be displaced by the medium, and the relative ease with which this can be performed is of considerable practical importance.

The surfaces of pigments and extenders vary in their wettability, some being readily wetted by water, that is they are hydrophilic, others being only poorly wetted by water but readily so by oils, that is, they are oleophilic.

In general, most pigments and extenders are found to be readily wetted by solvent soluble media, that is, they are oleophilic. However, there are many pigment-medium combinations that are difficult to disperse readily. Notable among these are carbon black and Prussian blue in solvent soluble media and most organic pigments in the aqueous solutions used in emulsion paint manufacture.

In these instances, the wetting process can be facilitated by the addition of pigment dispersing agents or, as they are alternatively known, surface active agents. Surface active agents are organic compounds which contain a polar (hydrophilic) and a non-polar (oleophilic) group in their structure. The modification of the pigment surface involves adsorption of either the polar (hydrophilic) or non-polar (oleophilic) group of the surface active agent molecule with the non-adsorbed portion extending away from the pigment surface. The group adsorbed depends upon the nature of the pigment, for example, pigments with hydrophilic surface characteristics preferentially absorb the polar or hydrophilic part of the surface active agent and vice versa. During this adsorption process, the gases and moisture present on the pigment are replaced by the surface active agent.

Typically, the dispersion of an oleophilic pigment in a medium such as water would be difficult due to poor wetting of the pigment by the dispersion medium. Addition of a suitable pigment dispersing agent will result in adsorption of the oleophilic (non-polar) portion of the molecule onto the pigment surface, whilst the hydrophilic (polar) portion of the molecule will 'project' out into the aqueous medium. Consequently, the oleophilic pigment surface will then be rendered hydrophilic due to the presence of the surface active agent and there will be readier wetting of the pigment by the medium.

Many types of surface active agents are used as pigment dispersing agents in paint manufacture. Compounds such as the long chain condensation products of ethylene oxide with fatty acids or amides, sulphonated oils and alkali metal soaps of carboxylic acids are used to assist dispersion of pigments in hydrophilic media. Short chain condensation products of ethylene oxide with fatty acids or amides and quaternary ammonium compounds are used to aid dispersion of pigments in oleophilic media. Addition levels of surface active agents are normally very low, additions of 0.1-0.5% on paint solids being typical.

4.6 PAINT VISCOSITY MODIFIERS

In most instances, when the basic raw materials comprising a paint (namely, pigments, extenders, resins and solvents) are dispersed together, the resultant composition would have little practical value. In particular, the viscosity (thickness) and flow properties would be incorrect and it is most likely that the resultant paint would be too thin. Consequently, it would be prone to heavy pigment settlement and have excessive tendency to flow, resulting in low film build during application. In many paint systems low viscosity is a desirable attribute, for example, with spraying lacquers and penetrating sealers and stains. However, with many systems such as decorative brush-applied paints and high build protective systems, a relatively high viscosity is a necessary requirement. In such instances, thickening agents are added to the paints during manufacture to increase the viscosity, although care has to be exercised to ensure that the increased viscosity does not adversely influence the flow and levelling characteristics of the coating.

The viscosity of many simple liquids such as water and organic solvents is independent of the rate at which the liquid is sheared, that is, they have an invariant viscosity at a given temperature. However, with all high molecular weight liquids and pigmented fluids the measured viscosity is dependent upon the rate at which shear stresses are applied to the liquid and, with such systems, increasing rates of shear result in greater fluid flow rates, that is, lower viscosities are found. A further factor of importance, particularly for brush applied finishing systems, is thixotropy. A fluid is said to be thixotropic if its viscosity varies with both the rate of shear and the duration of the shearing stress. In general, a thixotropic fluid will exhibit, at a given shear rate, a decreasing viscosity the longer the shear stress is applied to the system.

In practical terms, a thixotropic paint during in-can storage is subjected to zero or very low levels of shear stress and the paint will have a soft gel-like structure. On application, when the paint is subjected to high shear stresses, the gel breaks down and the paint becomes a mobile fluid and in this state, flow and levelling (movement under low shear rates) can occur. After application, the liquid paint will gradually regain a gel structure, although the strength of the regained gel is lower than that of its original, unsheared, precursor. The resumption of a gel state by the liquid paint is designed to occur before there is excessive flow of the applied paint, that is, before the liquid paint 'runs' away from sharp edges, or sagging occurs.

In contrast, the viscosity of a non-thixotropic paint is much less dependent on the shear rate and such paints remain liquid during storage. During application, when the shearing rate is increased, there will be an attendant reduction in viscosity which facilitates flow and levelling. After application, when the applied shear stress is removed, the viscosity of a non-thixotropic paint will revert to its slighly higher, initial value, but a gel state will not be reached unless there is very rapid solvent loss or other accelerated drying reaction after application. Consequently, there is a more ready flow of paint and this results in lower film builds with the possibility of sagging or running of thickly applied coatings.

There are many compounds capable of inducing high viscosity or thixotropy in paint systems and certain of the more important types are summarised below.

Natural Clays

A range of naturally occurring colloidal silicate clays known as bentonites have the property of swelling when dispersed in water. Therefore, they can be used as thickening agents in many types of water containing paints.

Bentonite clay, when treated with certain organic amine compounds, can be used as a thickening agent in solvent soluble paint systems. The treated bentonite clays are very light in colour and they are widely used in many types of coating systems. Addition levels are normally kept low, typically 2-4% on weight of paint, since high addition levels can result in the formation of excessively thixotropic systems with poor flow characteristics. They function by strongly absorbing polar solvent molecules onto their surface until eventually each particle is surrounded by a relatively large envelope of solvent, thus increasing the viscosity of the system. A gel is normally prepared by dispersing the treated clay in a non-polar solvent, e.g., an aliphatic hydrocarbon, which is introduced into the paint and then a small quantity of a polar solvent is added with stirring to cause gel formation within the system.

Thixotropic Resins

Reaction of an alkyd resin or an oil with a polyamide results in the formation of a resin with a thixotropic character. This type of resin can be used either as the sole binder in a paint or as an additive in conjunction with other, non-thixotropic, types of resin, depending on the degree of thixotropy required in the system.
The degree of thixotropy can also be varied within wide limits during the resin manufacturing process. The use of these thixotropic resins is common where thixotropy is required in brush applied finishing systems. One drawback to these resins, due to the presence of the polyamide component, is that they tend to yellow at a more rapid rate than conventional non-thixotropic alkyd resins.

Cellulose Ethers

Because of their swelling tendencies in water, modified cellulose ethers are widely used as thickeners in emulsion paint systems. In general, they do not impart thixotropy to such systems at their normal addition levels, 0.25-1.0% on the weight of the paint. Several types of cellulose ethers can be used in emulsion paints, two types commonly used being ethyl hydroxy ethyl cellulose and sodium carboxy methyl cellulose, the latter type imparting superior flow and levelling properties but inferior wash and scrub resistance compared with the former.

4.7 PIGMENT ANTI-SETTLING AGENTS

Many paints exhibit heavy settlement of the pigment component during prolonged storage, this particularly being so with pigments of very high specific gravity, or those with a very high degree of dispersion. The dispersion of settled particles prior to use of the paint is very time-consuming and can result in film 'bittiness' if not performed efficiently.
One way of reducing settlement is to impart thixotropy to the system so that the liquid paint gels on storage. Alternative methods are used where thixotropy is undesirable in the paint system.
Surface active agents, notably soya lecithin at levels of 1% of the pigment content, can be added to (solvent based) paints during the manufacturing process. The soya lecithin functions by adsorbing onto the pigment surface, increasing its volume and thereby reducing its density, so that the settling tendency is diminished (see Section 1.3).

A grade of calcium carbonate extender, surface coated with an organic compound, is frequently used in primers and undercoats, addition levels of approximately 5% of the pigment being typical. The extender particles have a relatively large volume due to the presence of the coating and they function by packing between the settled pigment particles, thus making it easier to disperse them in the paint medium.

5 Paint formulation

The formulation of a paint is largely determined by the properties required in both the liquid paint and its dry film. As has been previously stated, any paint can be considered as being composed of a pigment (including any necessary extender materials), a resin or binder and a solvent. The type of pigment and binder largely govern the properties of the resultant paint.

The properties and applications of the most commonly used pigments and binders have been considered in previous chapters. In this chapter it is intended to demonstrate how these materials can be combined to produce the many diverse types of paints and allied materials used in practice.

5.1 PRINCIPLES OF FORMULATION

Apart from the nature of the ingredients, a paint's properties are governed by the amount of each ingredient in the formulation. Most paint formulations are commonly considered, and expressed, on a percentage weight basis and, considering only the basic constituents, a paint formulation can be expressed as:

Pigmentation (including extenders) x% by weight
Binder solids (including plasticisers) y% by weight
Solvent (s) z% by weight
                                       ———
                                       100

From this mode of presentation, useful information can

be obtained which enables broad conclusions to be drawn concerning the likely performance and uses of the paint.

5.1.1 Pigment to Binder Ratio

The weight ratio of the pigment and extender content to that of the binder solids content, known as the pigment to binder ratio, can be used to classify paints according to their likely performance capabilities. Thus, paints with a very high pigment loading are not characterised by particularly good exterior durability. This effect is attributable to the inability of the relatively small amount of binder in the film to provide a completely continuous matrix for the relatively large amount of pigment particles present. Such films would erode rapidly on exposure. In general, paints with a pigment to binder ratio of greater than 4.0-4.5 : 1.0, irrespective of pigment or binder type, fall into this category and the prime use of such paints is for interior decorative applications with a low gloss finish requirement.

Paints with a low pigment to binder ratio, that is <1.0 : 1.0, are invariably high gloss finishing paints (although lower gloss levels can readily be achieved, if required, by the addition of small quantities of flatting agents) and these can be used on exterior as well as interior surfaces. At such pigment to binder ratios the influence of the constituents becomes more critical and this factor largely determines the end use characteristics of the coating.

Undercoats and primers, and certain speciality finishes, such as, exterior quality emulsion based paints and exterior flat alkyd finishes, typically have pigment to binder ratios in the range of 1.5-3.5 : 1.0.

5.1.2 Solids Content

The total solids content of a paint is also readily determined from a percentage weight formula. This figure represents all of the material that does not evaporate during establishment of the paint film on its substrate, that is, the pigments and binder solids. Accordingly, this solid material is frequently referred to as the non-volatile content of the paint.

The solids content of a paint is often independent of its viscosity, particularly if thixotropic agents are present, and as such is a useful check on paint 'quality'. In particular, it can be used to determine whether excessive amounts of thinners are being used during on-site application operations.

There are no well defined principles that dictate the solids content of paints and within any one class of materials, for example, primers and finishes etc., there can be a wide range of solids contents depending upon the nature and content of the raw materials used in the formulation. An indication of the solids contents of typical paint systems can, however, be obtained in Chapter 7.

5.1.3 Weight per Volume

A percentage weight formulation, together with a knowledge of the specific gravity of the individual components, also enables the weight per volume of the liquid paint to be determined. This figure is of great practical importance, since during the quality control procedures of paint manufacture significant deviations from the norm, representing addition errors, can quickly be spotted. This is particularly important during the manufacture of emulsion paints. Air entrainment due to incorrect milling procedures can result in short volume once the paint is filled out and stored in tins, because there will be a gradual escape of the air from the liquid. The weight per volume of the paint will detect this entrapped air and the fault can be remedied.

An example of a typical paint formulation, a white full-gloss alkyd finish, expressed on a percentage weight and a volume basis is shown in Table 5.1.

It should be noted that the resin in the paint formulation is used in the form of a solution in an appropriate solvent. The specific gravity shown in the table is that of the solution and not that of the solid resin component. During calculation of pigment volume concentrations (see below), however, the specific gravity of the pure solid resin, that is, independent of any solvents, is required.

A knowledge of the weight per volume of a paint enables the calculation of both applied film thicknesses and application rates.

Where the weight of paint applied to a known area has been determined, then the application rate, that is, surface area covered per unit volume of the coating, can be determined using the expression:

$$\text{Application rate, } m^2 \text{ litre}^{-1} = \frac{\text{Weight per litre of paint}}{\text{Weight per liquid paint applied to 1 } m^2}$$

(5.1)

	Weight, %	Specific gravity	Volume,
Titanium dioxide, rutile	27.0	4.2	6.43
Resin solution, 70% solids in white spirit, of a long oil length semi-drying oil alkyd resin.	60.0	0.93	64.52
White spirit	10.8	0.78	13.85
Calcium naphthanate, 4% metal content, 55% solids	1.0	0.91	1.1
Cobalt naphthenate, 6% metal content, 55% solids	0.34	0.98	0.35
Lead naphthenate, 24% metal content, 75% solids	0.86	1.3	0.66
	100.0		86.91

Theoretical weight per volume: 1.15 kg per litre
Theoretical solid content of paint: 70.4% by weight

TABLE 5.1   Comparison of Percentage Weight and Volume Expressions for a Paint Formulation

The wet film thickness of an applied paint can be obtained from the expression:

$$\text{Wet film thickness, } (\mu m) = \frac{\text{Volume of paint applied } (cm^3)}{\text{Area covered by that volume of paint } (m^2)}$$

(5.2)

In order to obtain the dry film thickness of the coating, this value for the wet film thickness has to be corrected for the volume solids content of the applied paint. This latter figure is derived from a knowledge of the mass and the density of the solid constituents of the formulation.

5.1.4 Pigment Volume Concentration (p.v.c.)

The specific gravities of the various materials used in paints, particularly those of the pigments and extenders (Table 1.3) and the resins (Table 5.2) vary over a wide range and this limits the usefulness of the information which can be obtained from a knowledge of only the pigment to binder weight ratio. This limitation is important when comparative performance tests are performed on paints of dissimilar compositions.

Binder types	Specific gravity
Vegetable oils	0.95
Alkyds	1.0
Chlorinated rubber	1.7
Vinyl copolymers	1.2
Melamine resins	1.2
Urea resins	1.25
Cellulose nitrate	1.7

TABLE 5.2   Average Specific Gravity of Various Binders

The volume concentrations of the constituents of paints, however, enable a more scientific approach to the interpretation of test results to be adopted. In practice, many physical properties of coatings are found to vary in a well defined fashion with the volume concentration of the pigmentation in the resin component of the dried film. This parameter is known as the pigment volume concentration (p.v.c.) and is defined by the expression:

$$\text{Pigment volume concentration (p.v.c.)} = \frac{\text{Volume of pigments and extenders}}{\text{Volume of pigments and extenders + volume of nonvolatile binder}} \times \frac{100}{1} \%$$

In paints formulated with a low p.v.c., there is an excess of binder present which results in a well bound film, and the nature of the binder has a dominant effect on the performance of the coating. Such paints would also be expected to have good weathering characteristics, probably a high gloss level and, depending on the nature of the binder, good chemical, water and abrasion resistance. An increase in the p.v.c. can only be accomodated by a relative decrease in the binder content and thus the influence of the type of pigment-extender becomes of greater importance. At extremely high p.v.c.s, there may be insufficient binder to firmly bind the particles together and such paints would be used primarily in interior situations where a high degree of wash and abrasion resistance would not be required. Also, as a consequence of the low binder content, such paints would have a low gloss level, poor flexibility and inferior chemical resistance characteristics.

Paints with intermediate p.v.c.s generally have properties somewhere in between these two performance extremes. However, with most pigment (extender) and binder systems many film properties are found to undergo sudden marked changes as a particular p.v.c. level is exceeded and, in general, these changes are not considered desirable in high performance systems. This level of pigmentation is known as the critical pigment volume concentration, c.p.v.c., and normally high durability coatings are formulated somewhere below this critical level. In contrast, certain types of lower performance coatings can be formulated above this level. In particular, this affords a cost advantage since pigments and extenders are cheaper than most types of binders.

The relationship between certain film properties and the pigment volume concentration is summarised in Table 5.3. The critical pigment volume concentration of paint systems is variable, depending largely on the nature of the constituents used in the formulation. In general, however, the critical level lies within the 45% - 55% p.v.c. range for many pigment-binder combinations.

5.2 METHODS OF PAINT FORMULATION

The effect of composition on the properties of paints has been briefly considered in the previous sections and it is clear that the ratios of the constituents in paints can modify the performance characteristics of coatings to a marked extent. Also of importance, however, is the nature of the constituents used for the preparation of a paint. Clearly, unwise selection could result in inferior performance. When formulating a novel paint, an indication of the likely service environment, life

Property	Low p.v.c. (<c.p.v.c.)	Change at c.p.v.c.	High p.v.c. (>c.p.v.c.)
Durability	high	marked	low
Permeability	low	marked	high
Blister resistance	low	marked	high
Gloss	high	not normally marked	low
Tensile strength	high	marked	low
Extensibility	high	not normally marked	low
Abrasion resistance	high	marked	low
Opacity	low	not normally marked	high
Cost of raw materials	high	-	low

TABLE 5.3   Comparative Paint Properties at Low and High Pigment Volume Concentrations (p.v.c.)

expectancy of the coating, method of application, colour, surface finish, drying time, and, necessarily, cost, is normally available. It is this knowledge, together with the availability of the constituents, that dictates the selection of the components of that particular coating.

The technique of paint formulation, that is, the bringing together of the individual constituents to provide a paint with all of the necessary performance and cost attributes required of it, generally involves a considerable amount of laboratory development work in order to achieve optimum results. Accordingly, it will be useful to consider here the techniques of paint formulations.

The performance characteristics of the many types of constituents that are used in paints are usually well documented by the manufacturers so that, to a large degree, it is possible to select the likely ingredients for a particular application simply by reference to the literature. However, the properties of these raw materials are frequently determined in isolation from other materials or, alternatively, in model formulations. Consequently, much of the original test data produced by the raw material manufacturer has to be rechecked in the particular formulation under consideration.

Paint type	Pigment to binder ratios
Primers:	
Aluminium for wood	0.7-0.9 : 1.0
Acrylic emulsion for wood	1.5-2.0 : 1.0
Lead-free for wood	2.0-3.0 : 1.0
Metal primers	2.0-4.0 : 1.0
Zinc rich primers	10.0-12.0 : 1.0
Undercoats: most types	2.0-3.5 : 1.0
Gloss finishes:	
Polyurethane - alkyd	0.85-0.95 : 1.0
Alkyd ⎫	
Chlorinated rubber	0.6-0.65 : 1.0
Amino-alkyd	(typical value, pigmentation
Vinyl copolymer (solution)	with organics involves
Vinyl copolymer (emulsion) ⎭	lower ratios)
Cellulose nitrate	0.2-0.5 : 1.0
Flat finishes:	
Alkyd, exterior quality	1.5-2.5 : 1.0
Emulsion, vinyl exterior quality	1.8-3.0 : 1.0
Emulsion, vinyl interior quality	4.0-6.5 : 1.0

TABLE 5.4   Typical Pigment to Binder Ratios of Certain Paints

For any particular formulation requirement, there are general formulation concepts that have become established through continued use. Typically, pigment to binder ratios (see Table 5.4) are frequently used to classify paints into broad groups, whereas the solids contents of paints are determined largely by the method of application required by the end user as well as on the basis of viscosity and storage stability requirements. Generally therefore, a request for a novel coating can be accommodated into a particular formulation category, for which the broad formulation ratios and the materials likely to be useful in such a formulation are established, so that development work can commence. Most paints generally have one or more prime performance requirements other than the normal characteristics required of all paints. These special properties are dictated by the service requirements of the particular paint. Thus a floor paint would necessarily need, in addition to the normal considerations of acceptable colour, opacity, gloss level and drying speed etc., the ability to withstand the high wear associated with floors.

During development work on a paint, special consideration is given to any such performance requirements, whilst trying to ensure that other performance attributes are not reduced to unacceptable levels. One common approach to optimising performance involves the use of a graphical technique whereby systematic changes in composition can be related to effects on one particular aspect of performance. This approach necessitates a knowledge of the important formulation variables that influence the particular property to be optimised and, in its simplest form, two formulation variables and their effects on paint performance can be studied with this technique.

As an example, assume that an interior quality white emulsion paint of low cost is required which needs as high a degree of wet scrub resistance as possible within the cost limits stipulated by the buyer. From a consideration of the general performance requirements of an interior emulsion paint, a combination of pigment, extender, emulsion resin, thickeners etc., can be selected. Clearly, a high degree of skill is required at this stage to ensure that suitable materials are selected. Correct selection will result in considerable savings in the overall development time.

A composition graph can then be constructed and, in an exercise designed to optimise the wash resistance of a paint film, it is likely that the pigment volume concentration and the amount of binder in the film will be of crucial importance. At a later stage, that is, once the ratios of the major constituents have been determined, the influence of minor components in the formulation could also be assessed using the following technique. The two composition variables would form the axes of the graph and the scale of the axes would be dependent upon prior knowledge of the performance, or likely performance, of the constituents under consideration. For example, a large scale length, covering as large a composition range as possible, would be used if there was little available data on the materials, whilst a shorter scale length could be adopted after either basic performance testing using the larger scale axes or if previous experience enabled certain assumptions to be made.

In use, three formulations are prepared from random points on the graph (Figure 5.1) selected to form a triangle abc. Again the lengths of the sides of the triangle chosen depend on the extent of previous experience. After preparation of paints corresponding to these formulation points, test panels are prepared and subjected to a quantifiable wet scrub test. From the test results, the formulation with the poorest performance, say c in Figure 5.1, is rejected. The mirror image of this formulation point, point d in Figure 5.2, is then prepared, tested in a similar manner and the result compared with that of formulations a and b. Again the formulation providing the poorest result is rejected and the whole procedure is continued until the optimum film wet scrub resistance is achieved. In practice, this is judged to occur when successive formulations circle one particular formulation point. However, some discretion is required in interpreting the results since the effect of the formulation variables on other performance attributes such as opacity, gloss, colour, cost, etc., might necessitate that a composition somewhat removed from that providing optimum scrub resistance be selected.

There are many other approaches to paint formulation, probably as many as there are paint technologists. The graphical approach described above, however, is commonly used and, with suitable modification to the technique, changes in more than two formulations variables can be assessed. A statistical approach to coating formulation development is also widely practised and this is of great value in assessing the influence of minor constituents on film or paint performance.

The above approaches are ideally suited for the formulation of novel paint products. However, much of the work of industrial and commercial paint laboratories is concerned with maintaining constant quality in established production formulations. The variable raw material supply position over the last few years has necessitated almost constant modification of production formulations in an attempt to maintain price structure and consistent quality in the face of shortages and raw material price increases. In this situation, comparative testing of paint formulations following the substitution of one or more constituents is often necessary. The procedure outlined above permits alternatives to be readily assessed although commercial

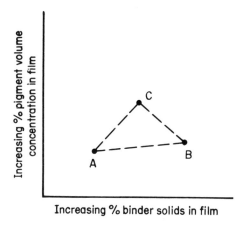

Figure 5.1   Schematic composition-performance graph

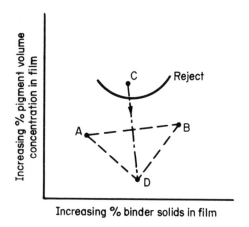

Figure 5.2   Schematic composition-performance graph

pressures are such that often only short term testing can be performed. Again the skill of the formulator is required in an attempt to extrapolate likely long term effects.

5.3 PAINT MANUFACTURE

The large scale production of paints is a complex process involving, for a pigmented system, dispersion of pigment and extender agglomerates into a resin solution, followed by relatively simple mixing operations to incorporate further liquid additions.

The important stage in the manufacturing process is the initial dispersion operation. Many types of machines are used to disperse the solid pigments and extenders into the resin component and the relative efficiency with which this process can be performed is of great practical importance. The speed with which pigment dispersion is achieved is largely governed by the type of machine being employed to effect dispersion and also by the rheological properties of the pigment-resin solution combination. In practice, the total pigment and extender content of the paint and sufficient of the resin is used, in the form of a low solids solution in the appropriate solvent, to produce a paste of sufficient mobility in the particular machine being used to effect dispersion.

There are many diverse types of equipment employed to disperse pigments in paint media. Traditionally, pug mills, edge-runners, roller mills and ball mills have been widely used, although with the exception of ball mills, the use of these types of machines is declining. Of more recent introduction are attrition mills and sand mills.

In a pug mill, a high consistency pigment-binder paste is subjected to mechanical breakdown by two intermeshing blades and the shearing action breaks down the agglomerates, allowing wetting by the binder. The dispersion efficiency is, however, rather poor and this class of machine is primarily used for the preparation of high viscosity fillers, putties and mastics, where a high degree of dispersion is not required.

A roller mill, generally consisting of three rollers, develops shearing stresses on a thin film of high consistency paste by virtue of the differential speeds of the rollers between which the paste is forced. As the rollers are very close together, this also imparts a crushing effect on the pigment agglomerates. Triple roller mills can be used for the preparation of paints requiring a relatively low degree of dispersion, such as primers and undercoats. However, they have largely been superseded in this function by ball mills. Single roller mills are still used and they are particularly useful for refining finished paint which may have developed a flocculate structure prior to filling out. With this type of roller mill the paint is sheared as it is forced, by the action of the roller, under an adjustable bar set at a close distance to the periphery of the roller.

An edge runner consists of either one or two heavy rollers revolving edgeways in a circular pan containing the pigment-binder paste. The weight of the rollers breaks down the pigment agglomerates, facilitating wetting by the binder. This machine is, however, rarely encountered in modern practice.

A ball mill is still the most widely used machine for effecting dispersion of pigments in solvent soluble media, although it is gradually being superseded by the sand mill. The mill consists of an enclosed cylindrical drum containing small balls which can be metal, sea washed pebbles or, more commonly, steatite. In use, the paste of pigment and binding medium is introduced into the drum with the balls, and the whole unit is revolved continuously. As the mill rotates, balls and paste are carried up the inside of the mill by centrifugal action, and cascade down when a certain height is reached, this height being determined by the speed of rotation. The breakdown of the pigment agglomerates in the ball mill occurs through the rubbing action on particles of pigment caught between the balls, and between the balls and the mill surface. The consistency of the paste is critical and it must allow the balls to move freely through it. In practice, dispersion times can be fairly short and typically a high gloss paint can be satisfactorily dispersed using an overnight dispersion stage. This leaves the working day to finish off the paint, drop it from the mill and to clean out the latter if necessary (i.e. for changing colour or product type), so that it can be recharged with a fresh batch of ingredients ready for overnight dispersion.

In an attrition mill, pigment agglomerates are broken down by imparting to them a high velocity by means of the rotation of a specially designed rotor blade in an open container. The pigment particles are forced outwards through the medium and are dispersed by a combination of the violent impact and shearing effect of the particles against each other and against the sides of the container. In order to obtain sufficient velocity, a very low viscosity medium is used. Attrition mills are widely used to disperse pigments during manufacture of water-based paints (i.e. emulsion paints). In this process, the pigments and extenders can be dispersed in the water phase with suitable additions of surface active agents. Once dispersed, the mill paste can be stabilised by the addition of the cellulosic thickeners. Using a much slower rate of revolution enables the shear sensitive emulsion polymer to be added together with the necessary minor constituents required to complete the paint formulation. Alternatively, after addition of the stabiliser, the paste can be pumped to a separate container for finishing, so that other dispersion operations can be performed in the mill. With this type of equipment a degree of dispersion adequate for an emulsion paint can be achieved in as little as 30 minutes.

Sand mills are of relatively recent introduction into the paint industry. They can conveniently be considered to be agitator-type ball mills. However, they are far smaller in size than the conventional ball mills and they have the further advantage of permitting dispersion on a continuous basis. The sand mill consists of a water cooled cylinder inside of which are a number of agitator blades which can generate rapid movement in the grinding elements. These grinding elements are usually spherical particles of quartz sand with a particle size of approximately 1 mm. In use, the sand mill can be mounted either vertically or horizontally, the pigment paste being pumped in at one end of the mill and forced through the grinding elements. The violent agitation of the sand induced by the agitator blades effects shearing of the pigment particles during their dwell-time within the cylinder. Dispersed paint is obtained from the other end of the cylinder through a screening device designed to retain the grinding elements.

With all methods of dispersion, optimum results are obtained by using only a relatively small proportion of the total binder requirement of the paint. After attaining the correct level of dispersion for the system, the remaining binder and any further liquid additives such as driers and solvents are added to finish off the paint. Where appropriate, the paint is then adjusted for colour by means of small amounts of liquid tinters, these being high solids dispersions of pure pigments in a compatible binder. This process is invariably necessary for coloured paints since, due to slight

variations in raw materials, addition errors and variations
in dispersion time, the batch-to-batch variation in colour
is often high. The colour matching procedure is a skilled
art, the object of which is to adjust the colour of the
dried film of the paint so that it exactly matches that
of an agreed standard. Concomitantly with the colour
matching process, the paint is subject to laboratory
quality control procedures as described in Chapter 9.

6 Techniques of paint application

The application of liquid paints to a substrate can be performed using a wide variety of techniques. Most commonly, brushes, rollers and spray guns are used, although there are many alternative methods such as dipping, roller coating and electrostatic spraying. As a generalisation, however, these latter techniques are more suited for the factory finishing of articles since complex and expensive equipment is required. The use of brushes and rollers is normally only encountered where on-site painting is required. Spray guns fill a compromise position, that is they can be used for both on-site and factory finishing.

During the paint formulation stage, due attention is always paid to the intended mode of application of the final product, since this determines the viscosity characterisitcs of the paint. In many instances, paints are supplied to users in a viscous state and require adjusting to a lower viscosity with suitable solvents prior to application using the desired technique. This has several advantages for paint users, notably problems of excessive pigment settlement on storage are largely obviated because the paint is diluted to its working viscosity immediately prior to use, whilst the user also has greater flexibility in the mode of application of his paint.

6.1   BRUSHES AND ROLLERS

6.1.1   Paint Brushes

Brushes are the most widely used tool for the on-site application of paints. They are relatively cheap, versatile in that they are able to cope with a wide variety of coating types, and they can often be used in areas where other application methods would be restricted. The best quality paint brushes are made from pure hog bristles which have a high paint holding capacity and provide a high standard of finish. Bristles, other than hog, are used in the manufacture of paint brushes and although none provide the quality of finish of hog, some types are ideally suited for applications where hog bristles would rapidly be damaged. Notable amongst these is the use of nylon brushes for painting rough surfaces such as concrete and masonry.

Paint brushes are available in a wide variety of sizes and designs to suit various types of paints and surfaces. Paints for brush application are normally fairly viscous and have a high solids content. The flow characteristics of brush applied paints need controlling with some care at the formulation stage and, typically, there has to be a sufficient degree of flow to enable the brush marks to level out, but not an excessive degree of flow such that the paint runs and sags on vertical surfaces or from sharp edges.

The drying times of brush applied paints normally have to be such that optimum flow can occur as well as maintaining the film in a state such that the overlapping of adjacent areas of freshly applied paintwork can be accomplished without film disruption along the interface. This latter effect is known as the 'wet edge' or 'open' time of a paint film and typically 5-10 minutes at normal working temperatures is considered to be acceptable.

6.1.2   Paint Rollers

The use of rollers for paint application saves a great deal of time, particularly where large flat areas are to be painted. They are particularly suited for the on-site application of emulsion paints and, as with brushes, rollers are available in a number of sizes and designs to suit differing areas. The composition of the roller can also be varied; mohair and lambs wool rollers both have a high paint holding capacity although they can impart a slight texture to the paint film, particularly with viscous emulsion paints. Synthetic foam-covered rollers are also available and these provide the smoothest surface finish but their paint holding capacity is lower than that of the natural fibre-based rollers. One limitation of rollers is their inability to cut into corners or work into confined spaces and it is normally necessary to use a brush for finishing off in such areas.

Paints for roller application are generally applied at a slightly lower viscosity than when applied by brush and, typically, this is achieved by the on-site addition of suitable solvents to a brushing quality paint.

6.2 SPRAY PAINTING

Although using simple and relatively low cost equipment, brush application is labour intensive and, in many instances where paints are applied on-site, the use of spray equipment, despite its high capital cost, becomes economically advantageous. Spray application of paints is, however, most widely practised in the factory coating of manufactured articles, sometimes termed industrial finishing. Here the benefits of spray painting lie in the speed, quality, economy and adaptability of the technique to accommodate most types of paint system and a very wide variety of sizes and shapes of the articles to be painted.

6.2.1 Conventional Spray Techniques

The equipment required for spray painting consists of a number of basic items namely:

(1) Spray gun

(2) Paint container

(3) Air hose

(4) Air transformer

(5) Air compressor and receiver

Certain other types of specialist spray equipment, notably the airless spray technique (see Section 6.2.3), utilise alternative technology.

The prime requirement for conventional spray painting is the provision of a supply of clean, dust-, water- and oil-free, compressed air. This is provided by an air compressor of suitable capacity to provide the volume and pressure of air required by the particular spray gun assembly and paint being sprayed.

A typical air compressor is of the piston type and there is usually an air receiver close by, which stores the air generated. The air receiver has a two-fold function. Firstly, it allows the air to cool (the process of compressing air raises its temperature) and thus to condense traces of any water vapour present and, secondly, the air receiver smooths out any supply fluctuations in the compressor and permits prolonged spraying with an even and steady supply of air.

Dust is eliminated from the air supply by the provision of suitable filters in the air intake of the compressor whilst water and oil in the air supply are removed by an air transformer. This device is a small, fully enclosed unit containing an array of baffles and filters. Air from the receiver passes through the transformer on its way to the gun and the impurities are removed by the filter units, provision being made at the bottom of the air transformer unit to enable periodic draining of the accumulated water and oil contaminants. An air pressure reducer is normally fitted to the air transformer unit and this enables the pressure from the compressor to be regulated to that required for spraying.

The compressed air is carried from the compressor to the spray gun by an air hose, and the length and internal bore of this hose is matched to that of the compressor-spray gun combination. This ensures that there is no undue air-pressure drop which would adversely affect the spraying operations.

In spray painting, compressed air and paint are supplied to the spray gun which, when triggered, allows the compressed air to force the paint through a fine orifice. This causes the paint to atomise, that is, form a fine mist of small liquid droplets. These droplets are propelled by the air stream to the article being coated where they are deposited and flow together to form a uniform film. The air pressure used in this operation is critical and it should be kept at the minimum required to atomise and deposit the paint onto the substrate. Too high an air pressure causes an excessive amount of the paint to rebound from the surface of the article being coated and this effect, known as spray fog, is wasteful of paint.

The spray gun is simple in principle. When a current of air is blown through a horizontal tube which passes over the top of an upright tube immersed in a liquid, there is a tendency to form a vacuum in the upright tube. This vacuum draws the liquid up the tube where it is entrained into the air stream passing along the horizontal tube. In design, commercial spray guns are precision-made complex pieces of equipment. The critical parts of a spray gun are those that are directly involved in the atomisation of the paint, namely the fluid tip, fluid needle and air cap.

The fluid tip is the small orifice through which the paint is forced on its transmission from the spray gun to the article being coated. The fluid needle is situated in the fluid tip and its function is to control the flow of paint from the gun. The fluid needle is actuated by a trigger set on the main body of the spray gun.

The air cap is the part of the gun that directs the air stream into the paint issuing from the fluid tip and causes it to atomise. The air cap also forms a spray pattern, that is, it controls the 'shape' of the stream of atomised paint from the gun, generally into either a circular or a fan pattern (as seen from the article being sprayed). On most guns, the spray pattern can be modified by an adjustor mounted on the body of the gun. The air supply to the air cap is also controlled by the trigger via an air valve device in the body of the gun.

When the spray gun is operated, the paint flows through the fluid tip and the rate of flow is dependent on the diameter of the orifice in the fluid tip as well as the air pressure and the viscosity of the paint supply to the gun. Correct selection and control of these various parameters is vital if paint is to be applied correctly and information on the required combinations of spray head, that is, air cap, fluid tip and needle, for various paint types is normally supplied by both the spray gun and the paint manufacturers.

The above design features are common to most types of conventional spray guns, but the mode of introduction of the paint into the gun is achieved by several different methods.

With the gravity feed gun, a small paint container is attached to the top of the gun body and the paint flows into the nozzle under the force of gravity whenever the trigger is activated.

Suction feed is another common mode of paint introduction. The paint container is attached to the body of the gun below the level of the nozzle and the liquid is pulled into the air stream through a tube extending into this container. With both the gravity feed and suction feed types of spray gun, paint container sizes of 0.5 or 1 litre are commonly used. Clearly, larger containers, if attached to the body of the gun, would be too awkward to handle by the operatives. Where large areas are to be painted such that frequent topping up of relatively small paint containers would be time-consuming and hence possibly uneconomical, the system known as pressure-feed paint spraying is widely used.

With this system, the paint container is a pressure feed tank, physically separated from the gun but connected via a fluid hose, having a capacity of between 2 to 200 litres. Air from the compressor is applied to the paint container, which is fully enclosed, and this forces the paint through the fluid hose to the spray gun where it is atomised in the conventional manner. The flow of paint from the container to the gun is dependent upon the paint viscosity, the length and internal diameter of the fluid hose and the pressure applied to the paint tank.

Irrespective of the spray gun design and the mode of paint introduction, the technique of spray painting is similar for all guns. The gun is held 20 to 30 cm away from, but at right angles to, the article to be coated, so that its contour is rigidly followed. The gun is moved at a steady, even rate over the workpiece which should be covered with the minimum number of strokes to avoid wastage of material. To ensure an even coverage, each stroke slightly overlaps its predecessor and, by releasing the trigger at the end of every stroke, overspray is reduced to a minimum.

Paints for spraying with the types of equipment described above are normally adjusted to a low viscosity (and hence low solids content) prior to spraying. Typically, solids contents of c. 30-40% and viscosities <1 poise (<100 $mN\ s\ m^{-2}$) are required for conventional spraying compared with the c. 70% solids and >3-4 poise (>0.3-0.4 $N\ s\ m^{-2}$) requirement of brush applied coatings. Consequently, the attainment of high film builds can only be achieved by the application of a number of coats although, by allowing several seconds for some of the solvent content to evaporate, these can be applied wet on wet.

All applied paints lose solvent as they dry, and due to the high price of most solvents, there would clearly be an economic advantage in avoiding the use of excessive quantities of solvents to adjust the application viscosity. Techniques for the spray application of high viscosity (and high solids content) paints have been developed and these can effect savings in time, since fewer coats are required to achieve the requisite film thickness. Furthermore, costs are reduced since the addition of large quantities of relatively expensive solvents is not required. Two such techniques are the airless spray and the hot spray methods, both of which can spray paints that would normally only be considered suitable for brush application.

6.2.2 Hot Spray Application

The hot spray technique of paint application involves the use of conventional spray equipment as described above.

A heater and heat exchange unit are, however, located near to the paint feed tank and paint is forced by a pressure feed system through this unit prior to being fed to the gun. To ensure that the paint is kept hot right up to the spray gun, a hot water jacketed fluid hose is incorporated in the system.

Paint temperatures in the range of 38-65°C are commonly used for spraying by this technique since over this range, the viscosity of most paint types is reduced by the same degree as that achieved by the addition of extra solvents for cold spraying. During the spraying operation, the rapid cooling of the atomised paint as it reaches the surface reduces any tendency to run or sag.

In addition to the obvious saving in solvents obtained with this technique, there is also a time saving since higher coating thicknesses are obtained more quickly than are possible with cold spraying methods.

6.2.3 Airless Spray

The airless spray method of paint application uses a different design concept to that of conventional paint spraying. The paint sprayed is atomised solely by projecting it through a special spray nozzle in the gun at pressures in the region of 20 MN m^{-2} (*c.* 3000 p.s.i.). The resultant pressure drop as the paint leaves the nozzle causes it to atomise. The very high pressures used in the airless spray technique permit nearly all paints to be sprayed in their original unthinned state.

The basic equipment required for airless spraying consists of a paint container, a high pressure fluid pump, a fluid hose and a special airless spray gun. In operation, the fluid pump, which can be either air or electrically driven, forces material out of the paint container at the required pressure. The paint passes through the high pressure hose to the spray gun where it is forced through a fine orifice in the spray cap. On emerging from the orifice, the paint expands and breaks up into a fine mist, and this atomised paint is forced into a preset spray pattern by angled slots in the spray cap.

Airless spray application produces a very wide spray pattern and, thus, it is ideally suited for the fast coverage of large areas. As there is no expanding compressed air stream to disperse the atomised paint particles, there is far less spray fog and consequently decreased paint wastage compared with conventional methods of spraying. This lower spray fog characteristic also makes airless spraying suitable for coating in enclosed

areas or wherever conventional spray painting would
be impractical.

6.2.4 Electrostatic Paint Spraying

The electrostatic method of paint spraying is a technique
designed for the automatic or semi-automatic coating of
articles on a conveyor system.

With this technique the paint receives an electric
charge, either from the gun which has to be of special
design or by passing it through an electric field after
being atomised. It is thus attracted to the earthed
article to be coated. As areas of the earthed article
are coated, they become electrically insulated and
further deposition can only occur at other uncoated areas.
This results in a coating of very uniform film thickness;
furthermore, spray fog and overspray are almost entirely
eliminated by this technique.

However, the process is limited by its inability to
completely coat internal or recessed surfaces due to the
shielding effect of the surrounding walls. Also, the
insulating effect of the paint film can make the application of more than one coat difficult. Consequently,
this technique is widely used for the application of one
coat finishes to domestic appliances and similar items.

Paints for spraying by the electrostatic process need
to have a similar viscosity to those applied by conventional
spray techniques. Further, the conductivity of the paint,
that is, its ability to carry an electric charge, needs
to be controlled within well defined limits, although most
types of oxidative and stoving cured single pack paints
can be applied by this technique.

6.3   DIP COATING TECHNIQUES

One of the limitations of conventional spray application
is that it is not suitable for the coating of articles
with complex shapes although, to an extent, electrostatic
spraying can accommodate such items. Furthermore, the
capital outlay and the degree of operator skill required
are both high.

Due to these considerations the process of dip coating
is widely practised and in many instances this has been
found to be the most economical method of applying paint.
As the name suggests, the technique of dip coating involves
completely immersing the article to be coated in a large
tank containing a quantity of the paint to be applied.
The article is then withdrawn, the surplus paint allowed

to run off into the tank, and the coating is either allowed to dry naturally or is force dried.

However, the art of dip coating is complex and requires careful planning to be successful. With articles of regular shape such as wooden kitchen implements and brush handles, which lend themselves to vertical dipping, there are few problems. One-coat finishes of good build without runs or sags can be achieved using air drying cellulosic or vinyl type lacquers. With articles of irregular shape, however, it is necessary to control the angle of dipping such that suitable drain-off points are obtained in order to eliminate excessive paint build-up at corners and complex angles. With such articles, special carrying jigs and semi- or fully automatic handling methods are usually adopted to perform the complete dipping operation. Many large articles such as motor car bodies can be coated on this principle, one of its advantages being that by complete immersion all exposed surfaces are coated. This technique is, therefore, ideally suited for metal primer application.

A low paint viscosity is a requirement for most types of articles to ensure an even coverage with good drain-off characteristics. Due to this low viscosity requirement, pigment settlement in the tank can be a problem. Small amounts of anti-settling agents are normally incorporated into the paint formulations which, together with mechanical agitators fitted into the dip tank, help to counteract the settlement process. Air drying paints can be used with this technique although, where rapid throughput and storage are required, stoving cured paints are preferred.

The main disadvantage of dip coating is the fire hazard involved in the storage of the liquid paint in the open vats (water based paints cannot be applied successfully with this technique). This problem is aggravated by the very large quantities of paint that are required, for example, many of the larger commercial plants have dip tanks with a 45,000-67,000 litre (*c.* 10,000-15,000 gallon) capacity. Because of the fire hazard associated with dip processes, alternative and less hazardous coating techniques involving full or partial immersion in the paint have been developed. Two such techniques are electrodeposition coating and flow coating.

6.3.1   Electrodeposition

The electrodeposition technique is an immersion process based on the movement of electrically charged particles of paint, under the influence of an electric field, to the workpiece which carries an opposite charge. Since the article has to carry an electric charge, the process is limited to the coating of metals. However, in order to carry an electric charge, the paints have to be water based and this effectively eliminates fire hazard.

The apparatus used for electrodeposition is essentially similar to that used for the dip coating of articles. Normally, the article to be coated is made the anode in a dip tank containing the paint. The tank itself and/or auxiliary electrodes placed in the paint, function as the cathodes and complete the electrical circuit. The aqueous based paint requires careful formulation and the resin component needs to be present in the form of ionised particles, typically, as the neutralised salt of an organic acid. During the paint manufacturing process, the (uncharged) pigment particles are wetted by the resin such that the composite carries the charge of the resin, which is normally negative.

During operation, current is passed through the cell causing the negatively charged paint particles to diffuse to the anodically polarised workpiece. At the anode, the paint is deposited onto the surface and this effectively insulates that area of the article from further deposition. Consequently, further deposition then occurs on the other areas of the article, the process continuing until all exposed areas are coated. On removal from the tank, surplus paint is removed by gentle rinsing with water prior to the stoving (baking) process required to cure this type of paint film.

The electrodeposition process is limited to one coat application due to the shielding and insulating effect of the deposited paint film. Its most common use industrially is in the application of primers to the chassis and body work of vehicles.

6.3.2   Flow or Flood Coating

Flow or flood coating is a deluging process widely used to coat very large articles or articles with highly complex shapes, for example, central heating radiators. The apparatus required is simple - a paint supply tank, typically, with a capacity of 90 litres ($c.$ 20 gallons), which also functions as a receiver for drain-off, a paint supply hose working from a pump and a paint spreader nozzle which may be fixed or hand directed. In use, the article is flooded with paint so that all areas are covered and surplus paint draining off the article is passed back to the supply tank for re-use.

Paints for application by flow coating are normally of low viscosity since the same considerations of drainage apply for flow coating as in dip coating. However, due to the complexity of the articles that normally require coating by this technique, paints based on slowly evaporating solvents, for example, white spirit, are normally used to ensure that the paint has an adequate flow time. The relatively small volume of paint requiring storage considerably reduces the fire risk as compared to the volume required for dip coating.

6.4 ROLLER COATING AND CURTAIN COATING

Roller and curtain coating are high speed techniques of applying paints to large flat panels prior to fabrication.

6.4.1 Roller Coating

A roller coating unit comprises a number of rollers arranged to take up paint from a feed tank and to apply it, at a controllable thickness, to the panel product which is passing between the roller assembly. These units are generally operated in conjunction with a stove oven unit. To ensure optimum flow, the viscosity of the paints used is normally low and the resultant film tends to be of relatively low build. Typically, metal and plastic foil and sheets are coated by this method.

6.4.2 Curtain Coating

Curtain coating involves passing the panel beneath a paint hopper, the base of which carries a slit, adjustable to suit both the width of the panel and the amount of paint required, that is, the coating thickness may be adjusted to a constant value for any given rate of panel throughput. With this technique, many types of paints can be applied to all types of metal, plastic and timber panels.

High film builds can be applied with the technique, which is unusual in that, with the exception of suitably modified conventional spray assemblies, it is the only method by which two pack paints, such as epoxies, polyurethanes and polyesters can be continuously applied, although twin feed hoppers and a suitable paint mixing head are required.

Clearly the field of paint application is very large and only a general summary has been presented here. For readers who require a more comprehensive coverage of coating techniques, a bibliography is included below.

6.5 BIBLIOGRAPHY

Industrial Paint Application, W.H. Tatton and E.W. Drew, Newnes, London (1964).

Paint Technology Manual Part Four: The Application of Surface Coatings, D.S. Newton (ed.), Chapman and Hall, London (1965).

Paint Finishing in Industry, A.A.B. Harvey, R. Draper, London (1958).

7 Substrates

The nature and condition of the substrate to which paint is applied is one of the major factors determining the durability of a coating system. Consequently, paints are usually formulated for use either on a specific substrate or a group of substrates. In this chapter, the paint formulations used for the three major classes of substrates of greatest interest to the paint technologist (metals, cementitious materials and timber) will be summarised.

7.1 METALS

Metalwork is classified under two broad headings, ferrous (steel and iron) and non-ferrous metals and, on economic grounds, steel is the most important substrate to the coatings technologist.

7.1.1 Steel

In order that paint coatings applied to steel substrates should retain their protective efficiency and aesthetic appeal, thorough surface preparation of the steel prior to paint application is vitally important if high coating durability is to be attained.

The widely used mild steels characteristically undergo a slow corrosion process (loosely termed 'rusting') on exposure to the atmosphere. The presence of certain contaminants such as chloride and sulphate ions as well as the gaseous oxides of sulphur and nitrogen, which are found in marine and industrial atmospheres, accelerates the corrosion rate of the metal.

A further characteristic of freshly fabricated steel is the presence on its surface of an oxide layer, known as mill-scale, formed during the high temperature processing stage. The presence of rust and/or mill-scale on the steel surface when it is painted is detrimental to the coating durability since neither substance can be considered to be a sound base for paint. If steel carrying such oxide layers is painted, underfilm corrosion and delamination of the oxide will occur, generally within a short time of exposure, and this results in premature paint system failure.

Many techniques have been devised for the removal of rust and mill-scale from steel, but the precise cleaning procedure specified for any steel article is dependent upon a number of factors. Typically, size and shape, access, the nature of the service conditions and economic considerations all have to be assessed before final selection of the appropriate cleaning technique.

7.1.1.1 *Cleaning Procedures*

A wide variety of cleaning techniques are in common use for steel, including wire brushing, flame cleaning, grit or shot blasting and acid pickling.

The most common method of cleaning steel is by the use of wire brushes and scrapers, either used by hand or power driven. This method, however conscientiously applied, is never completely effective as areas of mill-scale or rust are invariably left on the surface and become painted over. In many instances, however, scraping or brushing may be the only methods which can be used on-site, particularly where access is difficult. This method of cleaning never permits the highest durability in applied coatings to be attained.

Blast cleaning is a highly efficient mechanical method of cleaning mill-scale or rust from steel surfaces. The process involves impelling, at high velocity, a stream of shot or grit at the steel surface using a compressed air supply. The impact of the particles loosens and removes the oxide deposits leaving a clean, roughened, surface ideally suited for painting. The profile of the roughened surface can be altered within wide limits, from a smooth profile 'brush-off' finish to a rough profile 'white-metal' surface. With the coarser profiles, it is important to ensure an adequate paint coating thickness such that the peaks are not exposed.

Painting operations, or at least the application of a protective primer coat, have to be performed within 4 hours of blasting to ensure that surface rusting does not recommence. Blast cleaning is normally a factory process, although it can be used on-site. A variant of the technique is wet blasting. This involves impelling at high velocity a slurry of abrasive in water against the surface, and it can be used where the adherent oxide layers are contaminated by salts, as found under conditions of marine and industrial exposure.

Flame cleaning is a technique that involves directing an oxy-acetylene or similar hot flame onto the surface of the steel. Adherent oxide layers delaminate from the steel surface due to a different rate of thermal expansion, and can be removed subsequently by mechanical methods. In order to prevent condensation of moisture onto the substrate, painting should be performed whilst the steel is still warm. Flame cleaning is also useful for removing old paint layers, although the use of respirators is mandatory if lead-containing paints are suspected of being present on the metal.

Blast cleaning and flame cleaning are somewhat expensive procedures and are difficult to use for erected steelwork. However, the initial high cost is offset by the reduced maintenance demand, the life of suitable applied paint systems typically being 4-5 times that of paints applied to metal cleaned by wire brushing.

The use of pickling acids to dissolve and remove millscale and rust is a process normally only suited for factory application. In its simplest form, the article to be cleaned is totally immersed in a bath of hydrochloric or hot sulphuric acid which preferentially attacks and either delaminates or dissolves the adherent oxide layers. An inhibitor is normally present in the acid to reduce attack on the bare steel. With most processes, a phosphate and/or chromate film is deposited onto the surface of the steel directly after removal of corrosion products, and this is known as a pretreatment stage. After pretreatment, the final surface is an ideal base for paints, exhibiting improved paint adhesion and increased resistance to subsequent underfilm corrosion in service. Painting operations are normally performed within a short time of the pretreatment process in order to avoid atmospheric contamination and/or corrosion. This technique of acid cleaning followed by phosphate and/or chromate pretreatment is widely used in plants fabricating light industrial articles.

Acid cleaning by bath immersion is clearly impossible on erected steelwork. However, sprayable pickling agents are available which can be used on-site. They are normally only effective on light rust deposits and not on thicker mill-scale. In common with the factory processes, pretreatments such as phosphating can also be applied after the cleaning operations.

In addition to the removal of mill-scale and rust from steel, it is also necessary to ensure that surface contaminants such as oils and grease are removed. This is generally performed using solvent or alkali cleaners and, in practice, degreasing operations precede scale or rust removal by acid pickling but follow mechanical operations. However, surface degreasing after blast or flame cleaning is unnecessary since the surfaces produced are free of residual contaminants and can be painted immediately.

7.1.1.2 *Paints for Steel*

As a substrate, steel is presented for painting in numerous forms, including structural members for buildings, plates for marine applications, storage tanks, pressings for car bodies and home and office equipment. Selection of suitable paint systems for this diversity of end uses, each with their own specific performance requirements, is further complicated by the variety of service environments to which the article can be exposed.

The tendency for steel to corrode necessitates the use of coatings formulated to provide specific corrosion inhibiting properties. This characteristic is provided by the inclusion of certain inhibiting pigments in the primer coat (see Chapter 3), and normally all steel is coated with an inhibitive primer. There are certain exceptions, however, particularly where the expected exposure conditions are likely to be very mild, for example, with office equipment. Here, non-inhibitive primers or even one-coat finishing systems can be used although they are normally applied over some form of surface chemical pretreatment, such as a phosphate coating.

The atmospheric corrosion process of steel can be represented in a simplified form as:

(a) $4\,Fe \rightarrow 4Fe^{++} + 8e^-$ anodic reaction

(b) $4H_2O + 2O_2 + 8e^- \rightarrow 8OH^-$ cathodic reaction

followed by:

(c) $4Fe^{++} + 8OH^- \rightarrow 4Fe(OH)_2$

(d) $4Fe(OH)_2 + 2H_2O + O_2 \rightarrow 4Fe(OH)_3$ corrosion product

(7.1)

Paint films can provide protection by a number of mechanisms which may be summarised as follows:

(1) Suppression of the anodic reaction of the substrate (anodic polarisation)
(2) Suppresion of the cathodic reaction of the substrate (cathodic polarisation)
(3) Introduction of a high resistance into the circuit of the corrosion cell (resistance polarisation)
(4) Formation of a surface film which acts as a barrier against permeants

One or more of the above mechanisms can operate in the protection of steel by paint coatings. In practice, it is found that the nature of both the pigment and binder type used have a marked influence on the mechanism of protection of any particular formulation, and careful selection of the ingredients used in the various components of paint systems for steel is of great importance.

As previously stated, primers for steel are generally formulated to contain a corrosion inhibiting pigment, the function of which is to suppress (polarise) the corrosion process of the steel substrate.

Undercoats and finishing coats provide extra film thickness to the coating system, and specially formulated types can polarise reactions (b) and (d) in Equation 7.1. In many instances, compositional differences between the undercoat and finish are slight and it is common practice to have a colour difference between them to enable checks to be made on the quality and consistency of application. Undercoats and finishes are not normally pigmented with corrosion inhibitive pigments. Systems comprising primer, undercoat and finish are generally low build, $c.$ <100 µm, and as well as being protective have to provide a decorative function. This is achieved through conventional means, that is, by including pigmentation in the finishing coat to provide the required colour and surface finish. The binder of such systems is selected to withstand the expected service requirements of the article or structure.

Another approach adopted is to apply a high build system to the primed steel. These high build coatings are normally one coat, spray applied systems, and provide film thicknesses of the order of 100-200 µm or more. They are used where exposed conditions are likely to be severe and the aesthetic appeal of the coating is secondary to the protective value of the coating. The pigmentation of these

Property	Alkyd airdrying (medium oil-length)	Urethane (one-component) oil-modified	Urethane (two-component)
Ease of application	Excellent	Excellent	Poor
Chemical resistance	Poor	Fair	Excellent
Water resistance	Fair	Good	Excellent
Weathering resistance	Good	Good	Fair
Corrosion resistance	Fair	Fair	Excellent
Adhesion	Fair	Fair	Erratic
Impact resistance	Good	Good	Excellent
Flexibility	Good	Good	Excellent
Easy surface preparation	Fair	Fair	Poor
Toxicity	Good	Good	Poor
Mar resistance	Poor	Fair	Excellent
Minimum expected service life (years)	2-3	2-3	3-4

TABLE 7.1 Comparison of Standard Maintenance Coatings (From: International Finishing Industries Manual, 4th ed., Wheatland Journals Ltd. (1975) p.29, by kind permission of the editor).

Phenolic (oil-modified)	Vinyl solution	Acrylic	Epoxy ester	Chlorinated rubber	Epoxy (two-component amine cured)
Excellent	Good	Poor	Good	Fair	Fair
Fair	Good	Fair	Fair	Good	Excellent
Good	Excellent	Good	Good	Excellent	Excellent
Good	Excellent	Excellent	Poor	Excellent	Poor
Good	Good	Fair	Good	Good	Good
Good	Poor	Poor	Good	Good	Excellent
Good	Fair	Fair	Good	Fair	Poor
Good	Excellent	Fair	Good	Fair	Fair
Fair	Fair	Poor	Good	Poor	Poor
Good	Good	Good	Good	Good	Poor
Good	Good	Excellent	Fair	Excellent	Excellent
2-3	3-4	2-3	1-2	3-4	3-4

high-build finishes typically includes conventional (non-inhibiting) colouring materials as well as pigments and extenders that provide barrier action and hence suppress reactions (b) and (d) in Equation 7.1. The use of high durability binders is essential with such systems.

The mechanisms of protection and the uses of the protective pigments used in steel coatings have been summarised in Chapter 3. Performance details of the binders typically used in protective coatings are summarised in Table 7.1 and have been discussed in Chapter 2.

Some typical formulation examples of paints used to protect steel are given below.

Etch primers, which are also known as self-etch, acid etch, pretreatment, prefabrication or wash primers, are a widely used class of material. They can be considered to be intermediates between a factory applied metal pretreatment process and a full primer coat and are often used as temporary protectives prior to application of the complete paint system. Etch primers are based on a low viscosity vinyl solution pigmented with a low solubility chromate pigment. They can be applied by brush, spray or dipping and they function by both improving the adhesion of subsequently applied coatings and by providing a passive layer on the steel, they also reduce the risk of underfilm corrosion.

To achieve optimum performance, etch primers are applied at a rate to produce a dry film thickness of c. 12-15 µm, and curing can be achieved by air drying or, when factory applied, by subjection to an intermediate temperature, force drying regimen. Due to the low film thicknesses deposited, a further primer coat, generally a zinc chromate type (see later), is applied prior to finishing.

A typical formulation for the etch primer in its normal two pack form, (that is, the components are stored separately and are only mixed prior to use) is shown in Table 7.2

Several variants of the formula given in Table 7.2 are possible. In particular, one pack formulations, are available with improved storage stability - 12-18 months compared with the 8-12 hours of the two pack formulation (when mixed). The performance capabilities of the one pack etch primers, however, are generally considered to be inferior to those of the two pack types. The water resistance of the basic formula shown in Table 7.2 is not particularly high but it can be upgraded by including an alcohol soluble phenolic resin, although such an inclusion will affect other performance attributes.

Component		% by weight
Resin	Polyvinyl butyral resin	7.2
	Zinc tetroxychromate	7.0
	Talc	1.0
	Isopropyl alcohol	50.0
	Toluol	14.8
Etchant	85% Phosphoric acid	3.6
	Water	3.2
	Isopropyl alcohol	13.2

TABLE 7.2 Two Pack Etch Primer

Primers based on zinc chromate are available in several forms and such primers are widely used for structural steelwork, where air drying compositions based on alkyd and oleoresinous varnishes are preferred. In the factory-finishing of light industrial articles, stoving cured compositions are often favoured. Such stoving primers are generally based on alkyds, modified alkyds or alkyd-amino resin combinations.

In general, primer coatings containing zinc chromate provide a high degree of protection to ferrous metals although their performance is reduced in heavy industrial and marine atmospheres. The light colour of the pigment, however, enables it to be used in protective coating systems where decorative finishes with a pastel or white colour are to be applied and, furthermore, the low toxicity of the pigment enables it to be used in situations where the more traditional lead pigments are excluded.

General purpose primers with a lower cost, based on combinations of (non-inhibitive) red iron oxide and zinc chromate, typically in the weight ratio of 5-10:1 for red oxide to zinc chromate, are available. These primers are used in situations where the exposure conditions are likely to be mild.

Zinc chromate containing primers are typically formulated at pigment volume concentrations (p.v.c.s) of 35-45% with paint solids contents of $c.$ 60%.

Zinc rich primers are normally based on air drying binders which are non-reactive with the zinc metal. Typical binders are chlorinated rubber, isomerised rubber, two pack epoxy-polyamides, polystyrene, inorganic silicates and vinyl polymers.

These primers function by providing sacrificial protection to steel substrates under corrosive conditions, that is, the zinc corrodes preferentially to the steel. In such coatings, the zinc metal content of the dried film must be sufficiently high for particle to particle contact to occur. This normally necessitates dry film zinc contents of $c.$ 95%, and a typical formulation example is given in Table 7.3.

Component	*% by weight*
Zinc dust	83.9
Chlorinated rubber	2.65
Chlorinated paraffin	1.75
Xylene	11.7

TABLE 7.3   Zinc Rich Primer

These primers require well prepared surfaces in order to achieve optimum performance, and they may be applied by either spray or brush.

Finishes for steel are based on binders selected to withstand the conditions likely to be experienced in the proposed service environment. For exterior exposure under mild conditions, systems based on air drying, long oil length alkyds are typically used. Exposure to increasingly severe conditions, however, necessitates the use of more resistant binders and, typically, for exterior exposure, suitably plasticised chlorinated rubber is favoured.

A typical formula for a low build, air drying, white gloss finish based on chlorinated rubber is presented in Table 7.4.

Alkyd resins can be used as alternatives to the chlorinated paraffin plasticisers, and such formulations tend to be more flexible and have better weather resistance but have a lower chemical resistance than the chlorinated paraffin plasticised paints.

Component	% by weight
Titanium dioxide	17.0
Chlorinated rubber	20.0
Chlorinated paraffin	13.0
Xylene	40.0
White spirit	10.0

TABLE 7.4 Chemical Resistant White Gloss Finish

The presence of white spirit (which is not a solvent for chlorinated rubber) in the above formula facilitates brush application by slowing down the drying rate of the composition. It also reduces the solvating action of the top coat on preceding coats of the paint system. If the coating is to be applied only by spray application, then the white spirit could be replaced by extra xylene and, consequently, faster drying rates would be achieved.

A compatible undercoat for the above chlorinated rubber finish would be formulated along similar lines to a conventional alkyd type, namely, a p.v.c. in the region of 40-45% and a total solids content of 65-80%, with a combination of pigments and extender materials. In this instance, however, chemically inert extenders such as talc, barytes and mica would be used to obviate downgrading of the performance of the system.

A one coat, high build protective finish for use under severe exterior exposure conditions would be formulated, using a suitable pigment-extender-non-convertible resin system, to a p.v.c. of c. 30% at a total solids content of 65-75%. Such a finish would be applied by airless or hot spray techniques.

For the factory finishing of office equipment and domestic appliances, where a high degree of abrasion resistance and resistance to water and cleaning fluids would be required, stoving-cured, spray applied finishing systems are generally used. Such finishes can be used as one-coat finishes, although to achieve the highest standard of finish and performance, they can be applied over a combined primer-undercoat. Due to the low corrosive severity of the environment that is likely to be experienced in service, such primers can be based on non-inhibitive pigments and extenders. A typical formula is given in Table 7.5.

Component	% by weight
Titanium dioxide	14.8
Barytes	13.6
Talc	13.6
Melamine-epoxy resin, 60% solution in xylene	29.0
Xylene	25.5
n-Butanol	3.5

TABLE 7.5 Non-inhibitive White Stoving-cured Primer-undercoat

This paint would be spray applied using conventional equipment and cured by stoving at 120°C for 30 minutes. It would possess excellent adhesion characteristics combined with a high degree of hardness and chemical resistance.

A typical formula for a stoving-cured gloss finish is presented in Table 7.6.

Component	% by weight
Titanium dioxide	26.6
Non-yellowing short oil alkyd, 60% solution in xylene	44.4
Melamine resin, 60% solution in xylene	12.2
Xylene	9.0
n-Butanol	7.8

TABLE 7.6 Stoving-cured White Gloss Finish

This finish would be spray applied using conventional equipment, and could be used either as a one coat finish or over a suitable primer-undercoat. Curing would be achieved by stoving at 80°C for 30 minutes.
For the highest degree of chemical resistance and adhesion in finishing coats, binders containing phenolic and epoxy resins are used. Such systems require high temperature stoving regimens to effect cure and would, typically, be used in situations such as the interiors of chemical storage tanks, barrels and drums.

7.1.2 Non-ferrous Metals

Many types of non-ferrous metals are used in industry, amongst these being aluminium, copper, brass, zinc and zinc-coated (galvanised) steel, lead and tin plate. Non-ferrous metals are usually more resistant to corrosion in the atmosphere and many other types of environment than ferrous metals such as steel. However, coating systems are normally applied if either long term protection or protection against specific corrosive environments is required.
Much of the discussion on the preparation of steel prior to painting applies in principle to non-ferrous metals, and surfaces should be dry, free from dirt, grease and loose corrosion or paint products at the time of application of paint. Degreasing is particularly important when painting new non-ferrous metal surfaces since they are more likely to be contaminated with grease from their manufacturing operations than to carry corrosion products. Millscale, as found with ferrous metals, does not occur with non-ferrous metals, although surface slag can form during casting procedures.
Chemical pretreatment methods have also been developed for certain metals, notably zinc and aluminium, and many of these are based on acid chromate processes. Etch primers are also widely used to pretreat these two substrates.
Abrasion is sometimes required to remove loose paint or corrosion products from non-ferrous metals. However, blast cleaning techniques are seldom used since it is normally found that hand cleaning procedures, such as wire brushing, are adequate and do not cause excessive damage to the metal surface.
The selection of paint systems for non-ferrous metals follows a broadly similar pattern to that for other substrates in that selection is based on the anticipated service environment. For exterior exposure, the use of

an etch primer-zinc chromate primer combination is common for all metals. Finishing systems are normally of low build, that is, <100 μm, and are based on suitably resistant media pigmented with (non-inhibitive) colouring pigments to produce the required aesthetic appearance.

Clear (unpigmented) coatings are also used on certain non-ferrous metals in particular applications. Typically, brass and copper, exposed internally or externally, can be coated with polyurethane or alkyd varnishes in order to preserve their natural decorative value. Clear coatings are also required for coating the interiors of aluminium food, drink and pharmaceutical storage containers and the main requirements of these coatings are high adhesion, non-toxicity and resistance to specific chemicals in the foodstuffs. Vinyl copolymers are typically used in this application and suitably pigmented variants are also used as one-coat decorative finishes for can exteriors.

7.2 CEMENTITIOUS SUBSTRATES

The substrates of interest to the paint technologist in this category include concrete, sand-cement render and asbestos-cement panelling. Also of interest, although not containing cement, are gypsum plasters, but for convenience these materials will be considered in this section.

All of the above substrates have similar characteristics. At some stage in their fabrication, water is added in order to effect setting of the material into a solid. This 'water of construction' is frequently retained for long periods within the substrate and its presence at the time of painting can adversely affect film performance. This moisture content also activates the alkaline salts that are formed in all of these substrates during their setting processes. The resultant surface alkalinity can result in chemical attack or saponification of certain types of binders used in paints, notably oils and alkyds, and this attack results in loss of paint film integrity such that the system suffers a marked diminution in its resistance to washing, abrasion and weathering. In extreme instances, where the moisture content of the substrate is high, saponified paint can be flushed off the surface.

The cementitious substrates are more alkaline in nature than plaster and, as such, present greater problems when being painted. The possibility of alkali attack on dried paint films, however, diminishes as the substrate ages, owing to the drying out of the material itself and the progressive neutralisation of the alkalinity by reaction with atmospheric carbon dioxide. The painting of dry, aged, cementitious or plaster surfaces presents no problems and any type of paint system, both alkali and non-alkali resistant, can be used. However, the neutralisation of the alkaline salts within the substrate is essentially a long term process which may take several years to be completed for materials in thicker sections. Because of this, the accidental wetting of the dried substrate, particularly the cementitious types, can reactivate the salts with subsequent attack on sensitive paint films. Consequently, the application of alkali resistant paints is normal practice irrespective of the apparent age and condition of the substrate at the time of painting.

A further characteristic of these substrates is their porous nature. This porosity normally necessitates that primers used for these substrates should contain sufficient binder to satisfy the suction of the surface without leaving the film underbound. In many instances, low viscosity, unpigmented binder solutions are applied prior to the full paint system in order to counter any effects due to surface porosity. These materials are generally termed 'sealers' and they also have the characteristic of consolidating any friable matrix material present on the substrate surface.

Paints for cementitious substrates almost invariably are required to be suitable for exterior exposure since these surfaces are rarely encountered on the interiors of buildings. The reverse situation exists when considering plaster, since this substrate is exclusively used on interior situations. However, although exterior durability is not a prime requirement, the abrasion resistance, wash resistance and the normally high aesthetic requirement necessitates that high quality paints have to be used on plaster. Paints for both cementitious materials and plaster normally have a low gloss.

In formulating general purpose paints for these substrates, the use of emulsified binders, where the resin is in the form of small discrete particles dispersed in an aqueous phase, has become normal practice. Such paints possess good alkali resistance and weathering characteristics as well as having the advantage of being water dispersible. A further practical advantage of this formulation technique is that the convention of applying two or three paints of dissimilar composition, as for example, primer, undercoat and finish, has largely been superseded. In general, one composition is applied in a sufficient number of coats (normally two) to obtain the correct degree of film build.

Where exposure conditions are particularly severe, for example, in marine or heavy industrial environments, coating systems based on solvent-soluble binders, such as, chlorinated rubber or soluble vinyl resins, are normally preferred to the emulsion based coatings. This preference arises primarily from the superior water, chemical and weathering characterisitcs of the former. The formulation of such systems would follow that outlined in Section 7.1, but with the inhibitive primer coat being replaced by a sealer (see below).

The technique of formulating emulsion paints differs from that for solvent-soluble types. The resin used in emulsion paints is supplied as an aqueous dispersion, and a wide variety of resins is available in this form. Examples include polyvinyl acetate homopolymers plasticised with phthalates or copolymerised with various acrylic resins, pure acrylic copolymer resins and styrenated acrylics. All of these resins form films by evaporation of the aqueous phase and concomitantly with this process, the resin particles flow together or coalesce to form an integrated film. In practice, coalescence is temperature-dependent and in most systems, small additions of specific solvents are added to depress the temperature at which film formation will occur. These solvents are known as coalescing solvents and they function by softening the resin particles.

Various additives are required to be present in the paint, notably a thickener which, typically, is cellulosic in nature, a surface active agent to promote the pigment wetting process, and a biocide to inhibit growth of micro-organisms in the aqueous medium. The general properties of these materials have been discussed previously in Chapter 4. Other minor additives that are also added to emulsion paints include rust inhibitors, to obviate corrosion of the storage container, and anti-foam agents, which are surface active materials added to reduce frothing during the pigment dispersion process.

The presence within emulsion paint films of small quantities of water soluble materials (the cellulosic thickeners) can result in such films having a relatively low water resistance with possible adverse effects on film durability. However, by correct choice and use of the binder type, these effects can largely be minimised. The pigmentation of emulsion paints follows conventional practice except that for use on cementitious or plaster substrates, the colouring pigments have to be colour stable under the alkaline conditions prevailing at the substrate surface.

A typical example of a white emulsion paint (flat finish) for use on either interior or exterior surfaces is shown in Table 7.7.

Component	% by weight
Titanium dioxide	19.0
Whiting	22.0
Polyvinyl acetate-acrylic copolymer emulsion, 55% in water	28.0
Water	29.0
Cellulosic thickener	0.6
Pigment wetting agent	0.1
Biocide	0.3
Coalescing solvent	1.0

TABLE 7.7   Flat Finish White Emulsion Paint

This paint could be applied by brush, roller or airless spray techniques and two coats would be required on most surfaces. The drying time of emulsion paints is dependent upon the ambient atmospheric conditions. Two hours is typical under winter conditions or where air circulation is slow.

For exterior use, emulsion paints modified to have a fine texture are widely used (see Table 7.8), and the presence of the texturing agent, usually sand, provides greater durability through increased film-build and erosion resistance. These textured emulsion paints are also of value in levelling irregularities on the substrate surface and also in filling minor cracks. They do, however, possess greater dirt retention characteristics compared with smooth finishes and this can be a disadvantage in some situations.

Component	% by weight
Titanium dioxide	14.0
Talc	8.0
Whiting	14.0
Fine washed sand	20.0
Styrene-acrylate copolymer emulsion, 50% in water	25.0
Water	17.0
Cellulosic thickener	0.6
Pigment wetting agents	0.1
Biocide	0.3
Coalescing solvent	1.0

TABLE 7.8   Textured White Emulsion Paint

   The lower titanium dioxide content of these coatings is compensated for by the higher film builds obtained, whilst the need for higher pigment to binder ratios than for conventional emulsion paints is balanced by using an emulsion binder with an inherently higher degree of water resistance and pigment binding power.
   These types of textured finish paint are normally applied by brush or roller, although special types of gravity feed spray guns can also be used. Two coats are typical, although with extremely porous substrates and/or where the highest durability is required, these are applied over sealers. A typical sealer would be based on an alkali resistant, air drying resin. It is customary to use solution type resins such as chlorinated rubber, vinyl or hydrocarbon resins, since these, at the solids contents used (typically, 30-50%), possess good penetrating and binding ability. Pigmentation is not normally included in such sealers.
   The surface preparation required by cementitious and plaster substrates is minimal. Generally, it is sufficient to ensure that the surface is dry and free of loose matrix, dirt, moulds, etc. The loose matter can be removed by the use of a stiff broom and, where porosity is suspected or the surface appears friable, then such defects can be consolidated by the use of a sealer as described above.

Certain types of cementitious surface, notably cement render and plaster, exhibit a tendency to crack on ageing. These cracks are normally very fine and should be filled prior to painting. The fillers generally used for this application are highly filled materials with a minimum of solvent consistent with good working under the knife. Pigment contents are normally minimal but sufficient to impart the required colour. Relatively large amounts of extender materials are used to provide the required opacity (by dry hiding) as well as to control the working characteristics of the composition. A typical knifing filler (light grey) for cementitious and plaster substrates is given in Table 7.9.

Component	% by weight
Titanium dioxide	3.6
Carbon black	0.4
Barytes	15.0
Whiting	60.0
Short oil length oleoresinous varnish, 60% in white spirit	14.0
White spirit	7.0

TABLE 7.9   Filler for Cementitious Substrates

Due to the solvent content, such fillers are only suited for use with small cracks and surface defects, and their use for filling deep indentations or large cracks would result in the filler contracting away from the edges as the solvent evaporated. Consequently, where the capability to fill large defects is required, solvent-free binders should be used. Typically, two pack epoxy-polyamide solventless resins are employed, generally at somewhat lower pigment to binder ratios than shown above.

7.3   TIMBER

Timbers are normally classified into three groups, namely softwoods such as pine and spruce, hardwoods, typically oak and mahogany, and manufactured board materials such as plywood, chipboard and hardboard.

7.3.1 Interior Finishes

For interior use, timber requires painting in order to enhance and preserve the decorative and aesthetic appeal, as well as to improve its ability to withstand abrasion and wear. Exterior durability is obviously not a requirement for use in interior applications, although for the painting of softwood joinery etc., exterior quality paints are commonly used for convenience. Much of the timber used internally, for example, in veneer panelling for furniture,is in the form of plywood and this is invariably factory finished. Clear finishes are normally used in this application although opaque coloured finishes are employed for substrates such as chipboard and hardboard which have low inherent aesthetic appeal.

The finishing systems used in factory finishing of interior timber can be categorised into three groups depending on the resins used, cellulosic, polyester and alkyd-amino, with the last named having the greatest importance in the UK.

Cellulosic systems are based on nitrocellulose and film formation is by loss of solvent. A typical paint system would comprise a wood filler, a sealer to smooth the surface for subsequent coats and a finish which would be applied in a sufficient number of coats to obtain the required film build. The film build of cellulose finishes is generally low, and frequently three or even four coats are required to achieve a satisfactory appearance.

The wood filler would be formulated along similar lines to that for cementitious substrates (see Section 7.2) although the colour would be matched to that of the timber being filled. This normally necessitates the use of greater contents of colouring pigments. The sealer coat would be a low viscosity, low-solids variant of the cellulose finishing coat and, typically, it would contain 1-2% of a flatting agent, such as zinc stearate, to aid the sealing characteristics of the coating.

A typical gloss cellulose finish which would be spray-applied by conventional equipment is given in Table 7.10.

Polyester finishes are two pack catalytic-cured materials. They are based on oil-free polyester (alkyd) resins and styrene combinations, the reaction of which is catalysed by the addition of an organic peroxide. In this type of system, the styrene cross-links with the alkyd resin and,

Component	% by weight
Nitrocellulose, damped 30% with butanol	16.0
Short oil length non drying oil alkyd resin, 50% in xylene	10.0
Hard resin (fossil or ester gum)	10.0
Dibutyl phthalate	4.0
Methyl ethyl ketone	6.0
Methyl isobutyl ketone	16.0
Butanol	8.0
Toluol	30.0

TABLE 7.10 Cellulose Clear Gloss Finish

since no other solvent is present, the solid content is nearly 100%. This facilitates the application of high build, one coat finishes. A typical formula of a one coat clear polyester finish is given in Table 7.11.

	Component	% by weight
Resin	Polyester resin, oil free, 60% in styrene	80.0
	Wax solution, 1% in styrene	5.0
	Colloidal silica	1.0
	6% Cobalt naphthenate	0.5
	Styrene	11.5
Initiator	Methyl ethyl ketone peroxide	2.0

TABLE 7.11 One Coat Clear Polyester Finish

The function of the colloidal silica in the formulation of Table 7.11 is to reduce the sagging tendencies of the

applied coating. After application, the wax floats to
the film-atmosphere interface and reduces oxygen entry
into the film which otherwise would adversely affect
the curing reaction. After the film has cured, this waxy
layer has to be removed, normally by sanding, the final
gloss finish being achieved by buffing on polishing
machines. The reactivity of polyester finishes after
addition of the peroxide catalyst is very high and the
pot life of these systems is in the region of 15-30
minutes.

Methods to overcome this disadvantage for factory
application involve either the use of twin feed spray
guns or a curtain-coating twin-head system, with the
catalyst in one head and the resin and styrene in the
other. Alternatively, an 'active ground' system may be
employed, where the catalyst mixed with a nitrocellulose
coating is applied to the substrate. After drying of
the 'active ground' the polyester-styrene can be applied
and with this technique cure is effected by stoving at
a moderate temperature. The advantage of this approach
is that pot-life problems are entirely eliminated.

Alkyd-amino wood finishes are two pack catalytic
materials, with the curing process occurring only after
addition of the catalyst, which typically is a dilute
mineral acid. The pot-life of these materials runs into
several hours or even days, and air-drying times are
rapid (surface drying times, typically, are 20-30 minutes).
The drying time is, however, normally accelerated during
factory finishing by an intermediate temperature stoving
regimen. A formulation for a two pack alkyd-urea-formaldehyde
gloss finish is given in Table 7.12.

Component		% by weight
Resin	40% oil length non-yellowing alkyd, 50% in xylene	38.8
	Butylated urea-formaldehyde, 65% in xylene	45.0
	Xylene	9.8
	Methylated spirit	1.7
	Butanol	1.7
Initiator	20% HCl solution in methylated spirit	3.0

TABLE 7.12 Two Pack Alkyd-amino Clear Gloss Finish

This finish could be applied by spray or curtain coating and, generally, satisfactory one coat finishes can be achieved, although a sealer would be required on many substrates.

7.3.2 Exterior Finishes

The majority of timber used externally is in the form of solid timber, as opposed to laminates, veneers and panel products such as plywood. Both softwood and, to a lesser extent, hardwoods are used. Most painting is performed on-site, although it is common practice to factory apply priming coats to fabricated articles such as joinery prior to delivery to site.

A prime requirement of coating systems for externally exposed timber is the ability to protect the substrate from water and the damaging effects of ultra-violet light. Ultra-violet light is particularly harmful to timber, causing discolouration (generally bleaching) and defibrillation of the surface. The presence of water accelerates this degradation process but, more importantly, it causes dimensional instability of timber components. High moisture contents in certain non-durable timbers also lead to a decay risk.

Many pigments commonly used in paints, namely, titanium dioxide, carbon black and the iron oxides, have the ability to absorb, and dissipate harmlessly, ultra-violet radiation from sunlight. As such, they provide protection to both the binder components of the paint, many of which are susceptible to ultra-violet light degradation, and to the substrate. The use of clear varnishes is, however, common on many timbers, notably the tropical hardwoods, which have an inherently high aesthetic appeal and here the effective filtering of ultra-violet light is more complex. Many of the polymers used in their unpigmented form as timber varnishes have the ability to absorb ultra-violet light and although this initially protects the substrate, the absorption process by the polymer itself results in its own degradation with consequent rapid failure of the coating system. In order to protect varnishes from the damaging effects of ultra-violet light, it is necessary to incorporate absorbent materials, which are themselves transparent, in the polymer and, typically, complex organic compounds such as the benzotriazoles are used. A typical air drying gloss varnish suitable for exterior timber is shown in Table 7.13.

This varnish would be brush applied, in two or three coats at 16-20 hour intervals. To aid penetration into

Component	% by weight
Long oil length drying oil-modified alkyd resin, 70% in white spirit	85.0
Benzotriazole	2.0
White spirit	11.2
Cobalt naphthenate, 6% metal content	0.5
Lead naphthenate, 24% metal content	1.2
Methyl ethyl ketoxime (anti-skinning agent)	0.1

TABLE 7.13 Gloss Exterior Timber Varnish

the substrate, the first coat could be thinned prior to application with approximately 10% extra white spirit. Alternative binders to alkyds are frequently used in varnishes, notably polyurethane-modified alkyds and tung oil modified phenolic resins.

Pigmented finishing systems for exterior timber, and also for interior use on joinery, doors etc., are normally based on drying oil alkyd resins. Generally, a primer, undercoat and a finishing coat are applied, the latter invariably having a high gloss requirement. These materials are formulated for brush application, although much joinery delivered to site is factory primed, usually by dipping procedures. A general requirement for factory applied wood primers is that they dry quickly and this demand is satisfied by using short to medium oil length varnishes or alkyd binders. Force drying using stoving ovens is also used on many production lines.

Brush applied primers are not generally required to dry very quickly and these primers are normally based on the longer oil length alkyd resins. Acrylic copolymer emulsion resins are also used as primers (see Table 7.14) for timber and these do, however, possess more rapid drying times. Frequently, two coats of these emulsion primers are applied, the second coat acting as an undercoat.

Another widely used timber primer is the so-called aluminium primer, and this type is normally preferred on hardwoods, due to its penetrative nature and on resinous softwoods where the aluminium provides a barrier to resin exudation through the paint system. A typical formula is given in Table 7.15.

Component	% by weight
Titanium dioxide	20.0
Whiting	17.6
Acrylic copolymer emulsion resin, 50% in water	32.0
Water	28.6
Cellulose thickener	0.4
Pigment wetting agent	0.1
Biocide	0.3
Coalescing solvent	1.0

TABLE 7.14   White Acrylic Emulsion Primer-undercoat

Component	% by weight
Aluminium powder	18.0
Medium-long oil length tung oil phenolic varnish, 70% in white spirit	36.5
White Spirit	44.8
Cobalt naphthenate, 6% metal content	0.2
Lead naphthenate, 24% metal content	0.5

TABLE 7.15   Aluminium Primer

Representative formulations for a white alkyd undercoat and a white alkyd gloss finish are given in Table 7.16 and 7.17.
Prior to painting, timber surfaces should be clean and dry. The loose fibrous material found on weathered timber can be removed by sanding, either manually or by the use of a power sander, until a sound base is produced. Knots are generally either sealed with 'knotting' (a shellac resin solution in methylated spirit), an aluminium primer as described above, or alternatively cut out and replaced

Component	% by weight
Titanium dioxide	20.0
Whiting	17.0
Barytes	10.0
Talc	5.0
Long oil length drying oil alkyd, 70% in white spirit	36.0
White spirit	11.3
Cobalt naphthenate, 6% metal content	0.2
Lead naphthenate, 24% metal content	0.5

TABLE 7.16 White Alkyd Undercoat

Component	% by weight
Titanium dioxide	27.0
Long oil length, semi-drying oil alkyd resin, 70% in white spirit	60.0
White spirit	10.8
Calcium naphthenate, 4% metal content	1.0
Cobalt naphthenate, 6% metal content	0.34
Lead naphthenate, 24% metal content	0.86

TABLE 7.17 White Alkyd Gloss Finish

by a plug of sound wood. Cracks are filled prior to painting by using a filler of the type previously described, and surface mould and growths can be removed by washing the surface with a dilute solution of a household bleach.

From the brief summary presented here, it can be seen that many types of coating systems can be used on the various substrates that require paint either for decoration and/or protection. Selection of a particular coating system for a given substrate is dependent upon a number of factors, and it must be emphasised that the discussion presented here is not exhaustive. For those readers requiring more detailed information on substrates and appropriate paint systems, a bibliography of further reading is provided at the end of this chapter.

7.4 BIBLIOGRAPHY

British Standard Code of Practice, CP231; 1966, *Painting of Buildings*, British Standards Institute.

British Standard Code of Practice, CP2008;1966,*Protection of Iron and Steel Structures from Corrosion*, British Standards Institute.

Paint Technology Manual Part 1; Non-Convertible Coatings, A.H. Tawn (ed.), Chapman and Hall, London (1969).

Paint Technology Manual Part 3; Convertible Coatings, I.C.R. Bews (ed.), Chapman and Hall, London (1966).

Paint Technology Manual Part 4; The Application of Surface Coatings, D.S. Newton (ed.), Chapman and Hall, London (1966).

Protective Paint Coatings for Metals, J.A. von Fraunhofer and J. Boxall, Portcullis Press, Redhill, Surrey (1976).

8 Paint durability

The durability of a paint system is its ability to provide decoration and/or protection to its substrate over the period of time known as its effective service life. This period is the interval between the time of application of the paint system and the moment at which, through deterioration, it ceases to perform its required function. The durability of paint films is dependent upon a number of factors. These can be divided into 'internal' factors, such as formulation details and their effects on the physico-chemical characteristics of the coating, and 'external' factors such as the nature of the substrate, substrate preparation, location effects and weather. The latter two factors and their influence on paint durability will be considered in this chapter, the other factors having already been discussed in other sections of this book.

The environmental conditions to which paint films are exposed during service are of great importance in determining their performance. Knowledge of the anticipated service conditions is essential when formulating or specifying paint coatings. Environmental factors can be subdivided into the effects of location, that is, atmospheric pollution, and the effects of weather.

8.1   INFLUENCE OF LOCATION ON DURABILITY

The prevailing atmospheric conditions to which a paint film is subject can be categorised into four groups, namely, severe, moderate, mild and very mild.

Severe exposure conditions are experienced in coastal situations, within industrial regions, near to areas of heavy industrial pollution and where complete immersion in water or chemicals may occur. Paint systems required to withstand severe exposure have to possess the highest durability. Typically, this necessitates the use of paints based on chemical resistant binders such as the epoxide, polyurethane, vinyl or chlorinated rubber resins, and these materials are used in situations where limited access makes maintenance difficult or when cost considerations dictate a long maintenance-free service life. However, often conventional alkyd type paints are used, of necessity, in such severe exposure conditions, for example, where exterior joinery requires painting. The durability of these alkyd paints is not normally considered to be of the highest order but improved performance and an initial high degree of protection is conferred on the substrate by the provision of extra film thickness. This is most conveniently achieved by the application of two, rather than one, finishing coats over the undercoat and primer coats.

Moderate exposure conditions are experienced in light industrial and urban regions away from heavy industrial conurbations and coastal areas. Mild conditions are found in interiors subject to condensation, such as kitchens and bathrooms, or interiors where there is a source of pollution. Finally, very mild conditions are found in warm, dry interiors, not subject to pollution or condensation.

Paint systems for the last three exposure regions are, typically, based on alkyds, oleoresinous varnishes and emulsion resin binders. However, it is not uncommon for a paint system to be required to exhibit a greater durability than suggested by its immediate environment. Typically, interior paints which need to withstand abrasion or frequent washing down fall into this category, and here high durability paints would have to be specified.

The ability of paint films to withstand location effects is primarily dependent upon their chemical resistance, a characteristic dictated by the nature of the ingredients used in the formulation.

Most types of coating systems are resistant to the dilute acidic environments which can occur in certain exposure conditions, for example, in industrial or urban conurbations where sulphur-containing fossil fuels are burned. However, the presence of sulphur oxides in the atmosphere can cause increased drying times and premature loss of gloss in recently applied air drying, oxidative systems although, in general, these effects do not reduce the effectiveness of coating performance.

The formulation of paints resistant to concentrated acids presents a greater problem and most coating types are attacked on prolonged immersion in concentrated mineral and organic acids. Coatings for resistance to intermittent exposure to concentrated acids are generally formulated on chlorinated rubber or two pack epoxy resins.

Certain types of paints, in particular those based on oleoresinous varnishes and alkyds, are susceptible to attack in dilute alkaline conditions. Where alkali resistance is required, alternative resin types such as the polyurethane, chlorinated rubber and vinyl copolymers have to be used. In situations where air drying oxidative formulations are required, alkyds modified with these more resistant resins can be used to produce resistant paint systems. Resistance to concentrated alkalies, however, is more difficult to achieve since this medium tends to attack all coating types, particularly on prolonged immersion.

Coatings resistant to oils and organic solvents are also required in certain service situations and, as a generalisation, coatings with a complex cross-linked structure provide the highest resistance. Paint films can suffer severe attack from solvents, especially the chlorinated hydrocarbons (which are the basis of many paint strippers), ketones, esters and the aromatic hydrocarbons. Non-convertible coatings rarely provide good resistance to solvents.

The pigmentation of paints can also be adversely affected by pollutants, in particular, acids and alkalies. Susceptible pigment and extender types have been detailed in Chapter 3.

Oils and solvents generally have little chemical effect on pigments, although with certain organic pigments the phenomenon known as 'bleeding' can occur on overcoating. This effect is caused by solvents in the freshly applied coating solubilising the pigmentation in the underlying film. This solubilised pigment can be carried through the freshly applied film as it dries, usually resulting in a patchy surface discolouration. The phenomenon of 'bleeding' is especially prevalent where aromatic hydrocarbon, ester and ketone types of solvents are present in the paint formulation.

A further location effect requiring the use of specific coating systems is where resistance to high temperatures is required. Where temperatures up to $100^{\circ}C$ are encountered, as for example on central heating radiators or in the vicinity of certain domestic appliances, conven-

tional alkyd binders can be used. Modification of the
alkyd with the non-yellowing types of oil offsets pronounced
tendencies to film discolouration on exposure. Higher
temperatures, in the range of 100-150°C, cause progressive
film embrittlement with loss of adhesion in oxidative
alkyd paints and, consequently, paints based on polyurethanes or alkyd-amino stoving-cured binders are
favoured. These binder types also have little yellowing
tendency on prolonged exposure to these temperatures.

Resistance to temperatures in the region of 150-200°C
requires the use of paints based on silicone-modified
alkyds or the polyvinyl chloride fusion coatings of the
organosol and plastisol types. At these temperatures,
the colour of certain types of pigments can alter on
prolonged exposure and this necessitates the selection
and use of colour stable pigments. Typically, titanium
dioxide, carbon black and the cadmium pigments are colour
stable at elevated temperatures, whereas the iron oxides
and many organic pigments are subject to colour change.

Coatings required to resist temperatures in the region of
200-500°C are, typically, based on silicone resins with
both the organic and the inorganic types being suitable.
Pigmentation at these temperatures is difficult due
to colour stability problems, but in applications such
as industrial stacks and car exhaust systems, aluminium
pigment is favoured.

8.2 INFLUENCE OF WEATHER ON DURABILITY

The effects of weather on paint films is often marked
and can lead to rapid degradation of coating systems.
All of the components of weather - temperature, moisture
and radiation (sunlight) - can influence paint film
performance and, clearly, since weather is a complex
of these components, interactions can occur which render
particular combinations severe and others relatively
mild.

8.2.1 Resistance to Ultra-violet Radiation

Prolonged exposure of many types of paint films to sunlight results in their rapid degradation, and this is
primarily attributable to the ultra-violet wavelengths.
Ultra-violet degradation of paint films is a complex
process involving both an increase in the cross-link
density of the internal molecular structure of the
film binder and, concomitantly, a tendency for certain
constituent structural bonds to rupture. The result of
this process is that the film becomes tough during the

early stages of exposure. Eventually, however, the film becomes brittle, cracks, and ultimately, as water permeates through the film, it loses adhesion and delaminates (flakes) from the substrate. To offset this process, it is common practice to add pigmentary materials having the ability to absorb ultra-violet radiation to paints designed for exterior exposure. Titanium dioxide is an example of such a pigment and it functions by both absorbing the ultra-violet light and reflecting it away from the film, thus protecting the polymer. Invariably with this pigment, back-scatter from pigment particles in lower levels of the film results in a certain amount of polymer degradation and this process releases small amounts of the pigment from the matrix. This reaction, known as 'chalking', is exhibited to varying degrees by all types of titanium dioxide but it can be used to provide a degree of self-cleansing within coatings. It is, of course, detrimental to the performance of coloured paints containing titanium dioxide, since the colour of the film could alter markedly with prolonged exposure. However, by careful selection of pigments, this chalking process can be reduced to a minimum so that, for example, colour changes attributable to this mechanism will not be detectable for a period of several years.

Selection of the binder type is also important in determining the susceptibility of the paint film to ultra-violet light degradation. Polymers containing benzene ring structures are particularly susceptible to ultra-violet degradation since they strongly absorb in the ultra-violet wavelength region of sunlight, that is 290-350 nm. Binders without these structures are inherently more durable although, in many instances, other performance aspects may render them unsuitable for use in exterior situations.

8.2.2   Resistance to Moisture

Moisture also adversely affects paint film durability, particularly when it is present in the form of rain or condensation. The presence of moisture on a substrate during paint application is, generally, detrimental to paint durability since it will usually reduce the adhesion of the coating system. This effect is particularly marked on ferrous metal substrates, where underfilm corrosion processes can be initiated. With porous substrates such as timber and concrete, surface moisture, unless present in excess, is, generally, not detrimental to durability.

Condensation of atmospheric water vapour or rainfall onto the surface of freshly applied paints can effect a premature loss of gloss in all types of coating system due to disruption of the surface. Normally, however, this would not result in any long term reduction in paint durability.

Painting in conditions of high atmospheric humidity, that is greater that 90%, generally increases the drying times of air drying oxidative paint systems. However, the drying process is not entirely suppressed and provided that the film is not damaged mechanically or by prolonged precipitation of rain, snow, frost, etc. whilst in the wet condition, then subsequent durability is not likely to be affected.

8.2.3 Temperature Effects

The combination of low temperature and high humidity can also present drying problems, especially within the aqueous emulsion paint systems where film formation is due to evaporation of the aqueous phase and coalescence of the resin particles. With many emulsion systems, optimum coalescence will not occur at temperatures below c. $3^{\circ}C$ and since the film formed is not in a coherent state and will not subsequently become so, then the durability will be abnormally low.

Low ambient temperatures can also adversely affect the curing rate of many types of two pack epoxy- and polyurethane-based paint systems and certain of these coating types will not form films at temperatures below $10^{\circ}C$ without the use of additional catalysts.

Non-convertible coatings, that is, those that dry by solvent evaporation such as those based on chlorinated rubber and vinyl resins, will dry at very low temperatures and, as such, are ideally suited for winter use. However, the drying times of these types of coatings will be increased at any temperature in stagnant air conditions, since under these circumstances, the solvent released by the coating will tend to blanket the surface and impede the evaporation processes.

Tropical and high heat/high humidity conditions can also adversely affect the durability of paint coatings. Exposure to high air temperatures, that is, $50^{\circ}C$ or above, can result in a rapid embrittlement of paint films. In convertible systems, this would be due to an acceleration in the cross-linking rate, whereas with non-converting coatings, this could be due to a loss of plasticiser or residual solvent from the film. Furthermore, the expansion and contraction resulting from the temperature cycling of the substrate can induce cracking within brittle paint systems.

High atmospheric temperature effects can be reduced by the use of white pigmentation. Typically, surface film temperatures of paints with a white pigmentation are half those with a black pigmentation under conditions of summer exposure.

8.3   BIBLIOGRAPHY

Paint Film Defects, M.Hess, Blackwell, Oxford (1965).

Protective Paint Coatings for Metals, J.A. von Fraunhofer and J. Boxall, Portcullis Press, Redhill, Surrey (1976).

9 Testing of paints and paint films

In preceding chapters, the fundamentals of paint technology and the polymers, pigments and additives used in paints have been discussed. Furthermore, the formulation of paints and their application to substrates as well as the durability of paint films have been considered. The discussion of these topics in earlier chapters, however, either was based on published data on the various important parameters or indicated that it was possible to determine the factors which affect the suitability of a given paint for any particular service condition or application. In this chapter, the determination of the relevant parameters of paints as well as the evaluation of paint films are discussed.

It is important, however, to distinguish between paints and paint films or coatings. The generic term paint in the context of this chapter refers to the binder, pigments and other additives, that is, the liquid material which can undergo reaction or drying to form a solid continuous paint film. The paint film is the coating applied to a substrate through the application, and subsequent drying or curing, of a liquid paint to the substrate surface. Consequently, this chapter has been divided into two principal sections, namely, the testing of liquid paints and the characterisation of solid paint films. However, there is an intermediate stage between these two states, that is, the drying process of the liquid paint and the properties of the semi-dried film, and this is also discussed.

Standard test methods for most of the important properties of both paints and paint films are detailed in various National and International Standards. Furthermore, the basic theory and apparatus used in the test procedures for surface coatings have been discussed in some detail in the companion volumes to this work (see bibliography at the end of this chapter). Consequently, a generalised treatment of the evaluation of the more important properties of paints and paint films will be provided and the reader is advised to consult the standard specifications and reference works listed in Section 9.7 for the precise details of many of the test methods.

9.1   THE TESTING OF PAINTS

The various components contained in the liquid paint and the principles of paint formulation have been discussed earlier. Clearly, the liquid paint must contain a binder and its viscosity, which determines the flow characteristics of the paint, may be modified by addition of solvents. Similarly, the presence of pigments, extenders, thickeners, driers and other additives will affect the paint viscosity. The composition of the paint will clearly influence other characteristics, such as, flammability and flash point, as well as the setting behaviour and the properties of the dried film.

9.1.1   Viscosity and Flow

A prerequisite of any paint is that it should flow so that it can be applied to the substrate by brushing, flowing, dipping or spraying. Electrodeposited or electrophoretically deposited paints are unique in that flow *per se* is not involved in the paint application process. The parameter that provides the most precise indication of the behaviour of a liquid paint with regard to its application to a substrate is its viscosity. Various methods are used to determine the viscosity of paints, ranging from the simple, comparative flow cup tests to fairly sophisticated viscometric techniques.

When paint flows along a surface (or through a tube), a velocity gradient exists within the advancing paint front in that the outermost layer of the paint is moving far faster than the innermost layer adjacent to the surface. Thus a shear force or tangential stress is required to effect movement of the paint and the ratio of this tangential stress to the velocity gradient is a constant, known as the coefficient of dynamic viscosity. This coefficient of viscosity is commonly referred to as the viscosity of a liquid and it expresses the internal friction or resistance to movement of the liquid. If the coefficient of dynamic viscosity is divided by the density of the liquid, the kinematic viscosity is obtained.

There are, however, two principal types of fluid, Newtonian and non-Newtonian. Paints fall into the latter category. Newtonian fluids, when subjected to shear stress, exhibit flow that is directly proportional to the applied stress. In contrast, this proportionality does not exist with non-Newtonian fluids and the flow will vary in a non-linear manner with the applied stress. Thus with a Newtonian fluid, doubling the applied shear stress results in twice the amount of flow. Non-Newtonian fluids, however, may show an increase in flow that is more or less than this amount depending on the nature of the fluid. In fact, many paints possess the property of thixotropy in that a relatively high initial shear stress is required to start flow but thereafter flow occurs relatively easily with decreasing shear stress. Thus with non-Newtonian fluids, the apparent viscosity, which is the ratio of shear stress to the flow or shear rate, will vary with the shear rate.

Viscosity may be measured in several ways although only two basic approaches are involved. These entail measurement of flow through a capillary in which the rate of flow is proportional to the viscosity or measurement of the resistive force exerted by the liquid on a solid body moving through them.

The first method is utilised in the Ostwald viscometer, the Redwood viscometer and the various National and International Standard (B.S., ASTM, DIN, ISO) flow cups which are used to assess the 'consistency' of paints. In these methods, the time required for a given volume of paint to flow through an orifice or down a capillary is determined. From this determination, the viscosity may be calculated using basic physics principles or, more usually, by comparison with a suitable standard.

The alternative approach is to subject the paint to a known shear rate using a mechanically driven paddle and measure the force required to achieve this shear rate. A wide variety of commercial instruments are available which operate on this principle and they permit rapid and accurate determination of paint viscosities.

Clearly, viscosity measurement is important in the formulation of paints to ensure reproducible flow properties. The addition of solvents or thinners to modify the viscosity of a paint requires careful monitoring of the viscosity to ensure standardisation of different paint batches. The flow properties of wet paint films will be discussed in Section 9.2.1.

9.1.2 Density

The density or specific weight of a paint is an important characteristic in that it affects the thickness of the film formed by a given volume of liquid paint. Density is defined as the weight of a given volume of liquid. Often the specific gravity of a paint may be quoted and this is the ratio of the weight of a given volume of the paint to the weight of the same volume of pure water. Specific gravity is often the preferred quantity since it is expressed as a ratio, that is, a number, in contrast to density which is a weight per unit volume.

Density (or specific gravity) may be determined in several ways but commonly a pycnometer is used. This is a cup of accurately known volume which is filled with the test paint and weighed so that the density is obtained directly. Pycnometers are commercially available and 50 ml and 100 ml cups are most widely used.

9.1.3 Flammability and Flash Point

All paints containing volatile organic solvents have a certain temperature at which the solvent evaporating from the liquid paint will ignite in the presence of air when in contact with a flame or a spark. This temperature is known as the flash point and it is obviously of importance in determining the optimum methods of storage and transport of certain types of paint and solvents.

The flash point of a paint is dependent on the flash points of its constituent solvents, although other components (notably the binder which tends to raise the flash point) have a modifying effect. It is not normally possible to predict the flash point of paint, particularly those containing blends of solvents. However, paints containing a substantial amount of a highly volatile solvent, with a high evaporation rate, will, generally, have a lower flash point than those based on solvents of lower volatility.

The flash points of both the liquid paints and solvents are determined using either the Abel or the Pensky Martens testers. The Abel apparatus is designed for testing liquids with a flash point below $65^{\circ}C$ whilst the Pensky Martens is used for liquids with flash points above $65^{\circ}C$.

Both instruments operate in a similar fashion. A small quantity of the liquid under test is heated at a constant rate in an enclosed container. Periodically, a small flame is introduced at a pre-set distance from the liquid-air interface and, during this procedure, the container is momentarily opened so that the operative may observe the effects. When a particular vapour-air combination is achieved, the flame ignites the atmosphere within the container, exhibiting a small flash, and the temperature at which this occurs represents the flash point.

Many paints have flash points below the range of ordinary atmospheric temperatures and therefore can constitute a safety hazard. Accordingly, certain requirements have been established for the materials judged to be the greatest risks.

Under the provisions of the Petroleum Consolidation Act, 1928, materials containing petroleum products with a flash point below $23^{\circ}C$ are subject to certain storage and transport regulations. For the purposes of this Act, a petroleum product is defined as any derivative of coal, shale, oil, petroleum and bituminous substance.

Cellulose solutions, and flammable liquids for use with cellulose, with a flash point below $32.5^{\circ}C$ have to be stored and used in accordance with the Cellulose Solutions Regulations, 1938. For the purpose of this Act, a cellulose solution is defined as any solution of a cellulose derivative in a flammable liquid.

9.2 PROPERTIES OF WET PAINT FILMS

When first applied to a substrate, the paint flows to form a continuous surface coating. Thereafter, the paint is transformed from this mobile or fluid state, firstly, by solvent evaporation and then by a curing or drying reaction into a solid film. During the curing or drying process, the wet paint film should possess certain attributes as well as the ability to dry in a reasonable period of time.

9.2.1 Flow

It is clearly important that when paint is applied to a vertical or near vertical surface, it should remain in position and not exhibit marked 'sag', that is, flow under its own weight. If sagging does occur, then the final dried film will have an unattractive appearance whilst the substrate surface will not be uniformly covered.

Furthermore, the dried film might have unacceptably high internal stress levels arising from the film sag or from non-uniform drying rates resulting from the uneven film thickness on the substrate. Surface tension forces within the wet paint will tend to resist the sagging tendencies of the wet film but, for a given paint formulation, as the wet film thickness increases, the sagging tendency generally also increases. Since film sag can be controlled to a large extent by the additives present in the paint formulation (see Chapter 4), it is important that the sag propensity of a wet film should be measurable.

It is necessary, however, for a paint film to exhibit some degree of flow if it is to level and fill surface irregularities in the substrate. Consequently, some form of indexing or grading of flow is necessary to assess both the levelling and sag properties.

Sag properties may be assessed conveniently by means of a panel or block having a series of narrow, accurately machined, slots in its surface. Typically, slot depths ranging from 50 μm to 200 μm in 25 μm intervals are used. Paint is applied to the block such that the wet film thickness in each slot is half the slot width and then the block is placed on its side so that the strips of paint are horizontal. The sag index is taken to be the wet film thickness at which the width of the paint strip has increased by 25% due to sagging. This type of test is reasonably rapid, reproducible and simple.

A variation of this sag test incorporates a blade film applicator having several pairs of notches of increasing depth. The paint is applied to a flat surface which is then mounted vertically so that both the flow properties and the sagging tendencies of the paint can be evaluated by the character and the displacement of the applied film strips.

9.2.2 Assessment of Pigment Dispersion

The degree of pigment dispersion obtained during paint manufacture is of great practical importance. Inefficient dispersion results in poor opacity, poor gloss and colour development, as well as the possibility of film 'bittiness'. Conversely, an over-efficient degree of dispersion for the required end use of the coating involves unnecessary cost. The pigment particle size of a paint is known as the fineness of grind, since the pigment is dispersed and comminuted with the paint medium by grinding. Since this is a mechanical operation and the degree of dispersion and particle break-up is dependent upon the grinding conditions, it is necessary to be able to assess the fineness of grind. Various devices are available for this type of evaluation and these gauges are simple and rapid in use.

Fineness of grind is assessed in two principal ways, namely, by means of paint wedges produced between plate glass (e.g. the Garmson gauge) or with steel blocks that have wedge-shaped channels ground into their surfaces (e.g. the Hegman gauge). In the former, a paint wedge of specified thickness within the range 0-150 µm is produced between two pieces of plate glass. The pigment particle size may be ascertained by means of an inscribed scale at the edge of the plate glass pieces at the point where the diameter of the pigment particles exceeds the film thickness of the paint matrix. In the second test method, the gauge consists of a hardened steel plate with one or two grooves that, typically, vary in depth from zero to 25, 50 and 100 µm. The plate along the length of the groove is marked with the depth in µm and in Hegman units, the latter ranging from 0 (100 µm) to 8 (0 µm) in steps of 12.5 µm. The paint is drawn along the grooves by means of a straight edge and rippling of the paint film occurs at the point where the pigment particle diameter exceeds the groove depth. This type of gauge permits rapid and reasonably accurate assessment of particle size. Some gauges incorporate black and white test channels so the both pigment fineness of grind and the wet film opacity may be assessed. Paint opacity, however, will be discussed later in Section 9.5.3.

9.2.3 Wet Film Thickness

The wet film thickness of a paint coating often may have to be determined for a variety of reasons but any test method adopted is required to be rapid and reasonably accurate. Unfortunately, virtually no test for determining the thickness of wet paint films is non-destructive due to the extreme susceptibility to damage of the uncured coating.

In practice, different types of penetration depth gauge and the ASTM (American Society for the Testing of Materials) wheel gauges are most commonly used for wet films although a wide variety of techniques is available for dry films (see Section 9.3.1). The simplest types of depth gauge are combs which are available both as disposable plastic devices and re-usable precision ground steel or stainless steel gauges. The combs are produced

with the 'teeth' set to different heights so that when
the comb is pushed through the film and placed on the
substrate, the thickness of the film is read directly
from the 'tooth' that rests on the surface of the paint.
These combs are available to cover a range of film
thicknesses, typically, 0-120 µm, 0-400, 0-600 and
0-1200 µm, with from eight to ten teeth subdividing the
range of measurement. Such gauges are of low cost and
rapid in use but their accuracy is somewhat limited (±10%).

A rather more accurate type of gauge is the ASTM wet
film thickness gauge. This circular device consists
of three equally spaced wheels, the outer wheels being
identically sized, whilst the central wheel is smaller
and fixed to an eccentric cam. When the gauge is rolled
over a wet paint film, the eccentric central wheel run
picks up paint as soon as it comes into contact with the
film. The leading edge of the paint line on the central
wheel can be read from a scale on the outer wheel as the
wet paint film thickness. These gauges are available
for different measuring ranges, typically, 0-100, 0-200,
0-500, 0-1000 and 0-1500 µm as well as in a wide variety
of sub-ranges. The gauge is also rapid in use and is
reasonably accurate. Some of the more sophisticated
gauges discussed in Section 9.3.1 may be used, with
care and some modification, for the determination of
wet film thickness. In general, however, the high
degree of accuracy possible with such devices is not
required for wet film thickness determinations.

9.2.4 Drying Time

The rate at which a paint cures or dries is of major
technical and commercial importance for reasons such as
the delay necessary before re-coating or placing the
painted item into service. In testing the drying time,
however, three different types of determination are
performed, namely wet edge time, surface drying time and
hard drying time.

The wet edge time of a top coat paint is assessed by
the following procedure. A degreased burnished steel
panel (61 × 30 cm or larger) is provided with a compatible undercoat and allowed to dry for 24 hours. Then
one half of the panel is brush coated with the test paint,
the paint being applied parallel to the dividing line.
After a given time, e.g., 10 minutes, the second half of the
panel is painted, the paint again being applied or 'laid off'
parallel to the dividing line to join the two areas. After
drying, the joining line between the two painted areas is

examined and unsatisfactory, short wet edge time is indicated by imperfections at the join.

The surface drying time of a paint film is determined by coating a 15 × 10 cm panel of degreased, burnished steel, tin plated steel or aluminium with the test coating and allowing it to dry for a specified time. The panel is then positioned horizontally and silver sand (BS mesh size 52) is sprinkled onto the surface from a height of 15 cm. After approximately 60 seconds, the sand is brushed away with a camel hair brush and the surface is examined for retained sand or damage to the film. The surface is dry if no adherent sand is observed and no film damage is detectable. A variation of this test is that for tack freedom of the surface in which a small piece of gold leaf is placed on the surface under a weight of 800 g, for 10 seconds. The gold foil is then removed by tapping the panel and the paint film is free from tack if there is no gold attached to the surface.

The hard drying time of a paint is determined by a mechanical thumb test. A dry painted panel (15 x 10 cm approximately) is subjected to the action of a cloth-covered rubber tipped plunger that is loaded to 8.8 kg (the load may vary with the precise specification test). The plunger is rotated mechanically at a rate of 6 r.p.m. and, whilst rotating, is brought into contact with the surface and rotated through 270° (three quarters of a revolution) and then removed. The film has dried hard if the substrate, following testing, is not visible through the paint film, and no film damage has occurred. Various commercial instruments are available for performing this test so that it is possible to ensure a reproducible load, time of application, and test procedure, in all hard drying tests.

Tests for drying time are often performed in a modified form from those given above, typically trails of flock, bandage or sand can be applied to the surface or needles drawn through it. Furthermore, commercial and semi-automatic and automatic devices are available for these tests so that information can be obtained on all stages in the overall drying process of the paint.

9.3 GENERAL PROPERTIES OF DRY PAINT FILMS

The paint film on a substrate must fulfil a number of requirements if it is to be a satisfactory coating for the intended service of the painted object. Depending upon the service application of the object, the coating may have to be aesthetically pleasing, provide corrosion protection and exhibit satisfactory mechanical properties, as well as adhere to the substrate. These properties will be considered in the remainder of this chapter.

9.3.1 Film Thickness

The thickness of a coating is of major importance since it is this characteristic that profoundly affects most of the properties discussed later in this chapter. Furthermore, for any test on a paint coating to have relevance, it is mandatory for the film thickness to be known in order that the performances of different coatings can be accurately compared and assessed.

The determination of film thickness has been discussed in some detail in the companion volume to this work*. A brief resumé of the various techniques, however, will be provided here for completeness. There are in fact five principal methods of film thickness determination : weight change and dimensional change measurements, and the use of depth gauges, magnetic gauges and radiation meters.

Determination of paint film thickness by means of weight change measurements is less successful than for metallic coatings due to the considerably lower specific gravities of paint films compared to those of metals. Consequently, even for a thin metallic substrate, a very thick paint coating would have to be applied for an appreciable change of weight to be detected. Furthermore, the densities of most cured paints is not known so that an accurate calculation of the paint film thickness is often not possible even though the weight of the film may be determined with a reasonable degree of accuracy. The presence of two or three layers of paint, usually of different composition and pigment type and loading, complicates the matter further. Thus, although this approach is used in the laboratory and may be useful as a comparative test for coatings of the same composition, it is rarely applied on-site.

Dimensional change of a substrate following painting is, with suitably accurate vernier calipers or micrometers, an acceptable method of determining film thickness. This approach has the advantage of speed, non-destructiveness and repeatability. It is limited, however, by the accuracy of the measuring instrument and the operator skill, whilst thin films and irregular surfaces present problems of

Instrumentation in Metal Finishing, J.A. von Fraunhofer, Elek Science, London (1975).

measurement. An alternative approach which is useful for thin films and for multi-layered specimens is micro-sectioning. In this, the specimen, or a section of it, is mounted in an embedding medium such as acrylic or Bakelite resin (provided it can tolerate the curing temperature of these resins) or cold-curing epoxy resins. After embedding, the mount is ground back and the specimen is examined under a microscope. By this means, the total thickness of the coating as well as the thickness of the individual layers may be accurately determined. For very thin films, taper-section mounts may be prepared so that the film thickness can be magnified by the taper angle. Micro-sectioning has the advantage of producing permanent records (namely, the mounted section itself as well as photographs taken of the coating) but it has the disadvantage of being a destructive test in that the specimen has to be sectioned and mounted.

Depth gauges, such as combs (see Section 9.2.3) or feelers attached to dial gauges, are often used to measure film thicknesses. These gauges are essentially non-destructive in that there is minimal damage to the paint in testing although there is a degree of film disruption. Accurate depth gauges, which are more accurate than the comb devices, consist of rod or small anvil feelers attached to a dial gauge. The gauge is zeroed by placing it on the test surface and resting the feeler on the surface. The feeler is then pushed through the coating to the substrate and the dial reading noted. This reading is the film thickness. Alternatively, if the dial gauge cannot be set to zero, the difference in the two readings will give the film thickness. A wide variety of dial gauges are available and, depending on the range, thicknesses of up to, and greater than, 250 µm can be determined, often to an accuracy of ±2 µm.

Greater accuracy of measurement combined with true non-destructive testing may be achieved with magnetic gauges. There are, however, a variety of such devices which range from the very simple pull-off type to fairly sophisticated magnetic and eddy current gauges; the latter may be used with both magnetic (i.e. ferrous) and non-magnetic (non-ferrous) metallic substrates.

The simple pull-off gauge consists of a pencil-shaped barrel containing a small magnetic probe attached to a spring with a pointer that moves over a scale. The probe tip is contacted with the painted specimen and the barrel is lifted at right angles from the surface. The force required to detach the probe tip is inversely proportional to the thickness of the coating so that the scale on the barrel is calibrated directly as the coating thickness. The range of measurement is typically 0-600 µm. These devices are of reasonably low cost and rapid in use but suffer from limited accuracy, usually ±15%. Furthermore, they can only be used for non-magnetic coatings on magnetic substrates.

Rather greater accuracies (±10%) are possible with more sophisticated pull-off type gauges which incorporate spring or pivot beam balances. Since standard specimens are available, the force required to displace the gauge may be read directly as the coating thickness.

Accuracies of the order of ±5% are achieved with flux meters incorporating either permanent or electromagnets. These devices determine the proximity of a magnetic substrate by measuring the magnetic flux across an internal air gap. Since the coating determines the proximity of the ferrous substrate to the detector head, the coating thickness may be measured directly from an internal scale after calibration of the instrument. An advance on this technique is to use electromagnetic devices in which a small voltage (usually alternating) is applied to a yoke resting on the specimen. Since the coating functions as a virtual air gap, the voltage required to achieve an attractive force between the detector head and the substrate is related to the coating thickness, which may then be determined from a scale on the instrument.

Determination of the thickness of paint films on non-magnetic metallic substrates requires the use of eddy current gauges. The probe tip consists of a coil through which high frequency currents are passed. This induces eddy currents in the surface layer (the coating) and the magnitude of the induced or eddy current is proportional to the conductivity and the thickness of the coating so that direct measurement of the coating thickness is possible. Thicknesses of up to 2.5 mm on non-magnetic metallic substrates to an accuracy of ±5% or better are possible with this type of gauge.

In contrast to the magnetic gauges, the back-scattered beta-ray meter may be used for a very wide variety of coatings and substrates. The principle is that fast moving electrons or beta rays are emitted from a small radioactive source. When this radiation is aimed at a coated specimen, some of the beta rays will be reflected back by the coating whilst others will pass directly through the coating to the substrate where it may be absorbed or be reflected back. The proportion of radiation reflected back (back-scattered) varies with both

the coating and its thickness, since the intensity of
the back-scattered radiation will lie between the value
for the substrate and that for the coating. Calibration
of the instrument permits accurate determination of even
very thin coatings and meters are available for a wide
variety of coating/substrate combinations.

Finally, mention should be made of the optical paint
inspection gauge. This device is destructive in that a
cut is made in the coating and the total thickness
of the coating (as well as those of any component layers)
is determined by means of an internal microscope and an
optical vernier scale. These gauges are reasonably
accurate and have the great advantages of being usable
with almost any thickness of paint coating and any
substrate.

9.3.2 Adhesion

The adhesion of a coating to its substrate is clearly
of importance if the coating is to be retained under
service conditions. The phenomenon of adhesion is still
only poorly understood but it is thought to involve
mechanical interlocking as well as chemical bonding
between the coating and its substrate. Adhesion has been
the object of numerous studies and although it is
qualitatively and, to a degree, quantitatively understood,
there is no universal test method for measurement of the
adhesion between a coating and its substrate or even
that between two bonded bodies. Despite this limitation,
there are a number of test methods that are routinely
used to assess adhesion although, in general, the results
of such tests tend to be somewhat variable. Furthermore,
it is difficult to devise test methods which duplicate
the stresses imposed in service on the bond between
a coating and its substrate, notably dimensional changes
due to thermal effects, diffusion phenomena and, finally,
underfilm corrosion.

Four principal types of adhesion tests are in widespread usage, namely, mechanical deformation, peel,
scratch and pull-off tests. Mechanical deformation
tests fall into two broad categories, bend tests and
cupping tests. There are numerous examples of both,
as well as several standard specifications incorporating
such tests. There are two types of bend test in current
use, namely, those using cylindrical mandrels and those
requiring conical mandrels. The latter are more properly
used to evaluate the ductility or elongation properties
of coatings rather than their adhesion. A conical

mandrel bend tester consists of a 200 mm long mandrel which tapers from 38 mm to 3 mm over its length. The mandrel is incorporated into a framework into which the paint test specimen is clamped. The specimen is then bent over the mandrel by rotation of the roller frame of the tester. The elongation or ductility of the coating, expressed as a percentage, is obtained from a calibration curve using the minimum diameter on the mandrel at which failure (that is, cracking) of the coating was initiated. The cylindrical bend test requires a lever operated bend tester with either fixed or changeable cylindrical mandrels varying in size from the minimum diameter mandrel of 3 mm up to 25 mm or 32 mm in diameter. The test coatings are usually applied to 0.3 mm thick substrates of anodised aluminium or tin plated steel. The painted specimen is then placed in the tester having the largest diameter mandrel and is rapidly bent through $180°$ over the mandrel. If no failure occurs, other test specimens are bent over mandrels of decreasing diameter until cracks appear. This test permits a qualitative/semi-quantitative evaluation of coating adhesion as well as coating ductility. When performed under the conditions given in the various national and international standard specifications, this bend test represents a rapid, convenient and reasonably reproducible test method.

The other type of mechanical deformation test is the cupping test. In this, the painted test specimen is clamped in a die, and a punch, having a hemispherical head, is forced into the specimen, that is, the specimen is drawn, until failure of the coating occurs. The punch is hydraulically driven at a rate of 12 mm min^{-1} and at the point of failure, observed visually or under an optical microscope throughout the test, the drawing process is stopped. The depth of draw, recorded by a dial gauge reading to ±0.1 mm, gives the cupping value for the coating. This test is often known as the Erichsen cupping test and special cupping testers as well as auxiliary testing heads for tensile testing machines (see later) are commercially available.

Peel tests are often used to assess coating adhesion but such tests tend to be both qualitative and poorly reproducible. In such tests, one end of the coating is unattached or free of the substrate such that it may be folded back from the specimen. The specimen itself is then clamped in one set of jaws of a tensile testing machine whilst the tab or free end of the coating is held in the other set of jaws. The force required to peel off the coating is a measure of the adhesion of the coating to its substrate.

Adhesion may also be assessed by scratch testers although scratch tests should more properly be regarded as combined shear/scratch tests. In operation, scratch tests appear to be simple in that they involve drawing a needle or other cutting instrument across the coated specimen and determining the load on the needle required for complete penetration of the coating and contact to be made with the substrate. In fact, scratch tests are somewhat complex since shearing forces are applied by the loaded needle to the coated substrate and three parameters of the coating material can be assessed. Firstly, the load on the needle required to penetrate the coating is related to the adhesion of the coating to its substrate. Consequently, this test is a useful and a reproducible adhesion test. Secondly, the width of the cut produced by the shearing action is a function of the load on the needle for a given needle dimension and the shear resistance of the coating is obtained by dividing the applied load by the width of the cut. Finally, the scratch resistance of the coating may be assessed by determining the minimum load applied to the cutting head, which initially just rests on the coating surface, for it to penetrate the surface.

Various manual and automatic scratch testers are commercially available, the accuracy of determination and overall reproducibility of scratch tests being greater with the automatic testers. The great advantages of scratch testers are their relative simplicity and speed, whilst providing information on the adhesion, the cohesive and shear resistance properties of a coating as well as its scratch resistance.

Adhesion may also be assessed by a pull-off method. In this, a 'dolly' or test block is bonded to the coating on its substrate by a suitable cement or adhesive. The 'dolly' is then pulled from the specimen and the force required to detach the 'dolly' and the coating from the substrate is a measure of the adhesion of the coating to its substrate. Pull-off tests are useful in that a direct tensile load, without a shear component, is applied to the coating/substrate interface. The detaching force on the dolly/coating may be applied pneumatically or by a calibrated mechanical lifting device. A variation on the pull-off test, which also involves cementing or bonding a cylindrical test 'dolly' to the specimen, subjects the cylinder to a torsional force. The coating/substrate interface is subjected to a torsional force whilst the applied torque is linearly increased. A commercial instrument is available for performing this type of test automatically and the instrument incorporates a scale for measuring the stress at adhesive/cohesive failure of the coating and a scale for recording the axial torsional strain at failure. Consequently, the adhesion and cohesive characteristics of the coating may be assessed simultaneously. Torsional tests are also rapid and reasonably reproducible. It should be noted, however, that as yet no wholly satisfactory adhesion test has yet been developed although the tests referred to here do permit qualitative and, in some cases, semi-quantitative measurements to be made.

9.4 MECHANICAL PROPERTIES

The mechanical properties of a paint are clearly of great importance if the coating is to be able to withstand any degree of mechanical damage or even stresses resulting from dimensional changes due to environmental or thermal effects on the paint itself or its substrate. Consequently, various tests have been developed to assess the mechanical properties of paints and certain of these have been mentioned already in earlier sections of this chapter.

9.4.1 Tensile Strength

The tensile strength (or, more correctly, the ultimate tensile strength) of a material is the maximum stress or load per unit cross-sectional area that it will withstand before failure. It is determined by subjecting a specimen of the material, usually a flat bar or a rod of known dimensions and often of a particular or necked shape, to a tensile force in a tensile testing machine or tensometer. Many such testing machines are available, which range from relatively simple mechanical machines to very sophisticated electronic devices incorporating precision control mechanisms and recorders. Tensile testing using such devices is reasonably straightforward in that the test specimen is clamped between two sets of jaws. One set is stationary, being connected to a force balance or a transducer, whilst the other set of jaws is attached to a driving mechanism so that the two sets of jaws can be separated at a known and reproducible rate. This will subject the test specimen to a tensile load and it is possible to measure the force applied to the specimen as well as the displacement of the movable jaws, that is, the strain of the specimen. The latter measurement will be referred to again in Section 9.4.2 under ductility.

Modern tensile testers permit very wide ranges of strain rates and high test loads to be applied to specimens. In general, test specimens are necked or shaped so that the central portion of the specimen is narrower in cross-section than the ends. This ensures that failure always occurs within the necked section, which is very important for determining the ductility of the specimen. Furthermore, this procedure affords a means of testing very strong materials with testers of limited loading capability.

The tensile strengths of paints are usually determined by subjecting free films to tensile loading. The low strengths of paint films necessitates the use of sensitive tensile testers to ensure that meaningful data is obtained.

9.4.2 Ductility

The ductility of a material is its ability to withstand deformation without failure although it is usually specified as the deformation per unit length, that is, the strain, at failure under tensile or torsional loading.

The ductility of paints on metallic substrates is commonly assessed by 180° bend tests over cylindrical mandrels (see Section 9.3.2). In such tests, the mandrel diameter is decreased until cracking of the paint is observed. If the smallest mandrel over which the specimen can be bent without cracking has a radius of curvature r and the total thickness of the substrate and paint film is t, the ductility is given by:

$$\%\text{Elongation} = \frac{100t}{2r} + t \qquad (9.1)$$

There are various modifications of the standard cylindrical mandrel bend test such as the conical mandrel test mentioned previously and the Edwards test. In the latter, the coated specimen is bent over a standard G-shaped mandrel having a continuously varying radius of curvature on which a scale is scribed. The radius at which cracking is initiated is read directly from the scale and the ductility or elongation can be calculated.

Paint ductility can also be determined by applying it to a standard tensile test specimen and subjecting it to a tensile load. The elongation or strain at which failure of the coating occurs, denoted by edge cracks, can then be determined directly.

Cupping tests are frequently used to assess paint ductility as well as adhesion (see Section 9.3.2) and the ductility of the paint or other coating is expressed as a cupping value. Higher values always reflect greater ductilities. The advantage of cupping tests is that the specimen is subjected to simultaneous bilateral bending and stretching in all directions. Consequently, examination of the cracking pattern will indicate whether the ductility is uniform over the entire coating or if variations in ductility exist. It is difficult, however, to make quantitative determinations of paint ductility with such tests.

It is also possible to assess paint ductility from impact tests and these are considered in the next section.

9.4.3   Impact Resistance

Resistance of a paint to impact is an obvious requirement under most service conditions and, consequently, many standards and specifications prescribe testing of this property. In its simplest form, impact testing involves a free falling indenter striking the painted specimen which is then visually examined for surface damage. In fact, there are various types of impact but these tests fall into three major categories, namely, falling indenters, pendulum tests and particle jet tests.

The classic Izod and Charpy impact tests which are of great importance in the metallurgical and engineering fields are not used directly for paints, but a simplified form of pendulum test is frequently specified in various standards. In this, the paint is applied to two pieces of steel tubing (each 12.7 cm long and 3.8 cm in diameter) which are then attached to arms which swing in a vertical plane but at right angles to each other. The arms are then released so that the two specimens strike each other at $90°$ with a glancing impact, but ratchets ensure that they do not contact each other again on the return swing. After impact, the paint surfaces are inspected for damage. The severity of impact is controlled by the length of the arms and the vertical height from which the arms are released. The test represents a reasonably accurate method for assessing the impact resistance of coatings.

The alternative approach to pendulum testing is the free-falling indenter test. The indenter is usually a block or punch that has a rounded or hemispherical end and is of known weight. It is allowed to fall from a predetermined height so that it exerts a known force on the specimen which is placed on a die beneath the indenter. After impact, the nature and extent of any damage to the coating surface is assessed visually or microscopically. Testers are available having indenters of various weights and sizes with both fixed and adjustable dropping heights so that a range of impact forces are possible. Light duty testers which exert impact forces of 2-5 lbf in (0.23- 0.56 N m) are generally used for paint coatings but heavier duty instruments, having impact forces of up to 320 lbf in (36.16 N m), are used for plastics and electrodeposited coatings.

Commercial impact testers vary from the comparatively simple manually released indenter instruments to the more sophisticated mechanically and electrically released indenter devices.

Flexibility may be assessed by allowing the indenter to fall onto wedge-shaped specimens, that is, specimens that have been bent double over a mandrel and then are inserted into a hinged holder. The indenter falls onto the upper part of the hinge and forces the specimens into a conical shape. Alternatively, if indenters having ends into which several (usually 3 or 5) hemispherical knobs of different radii are embedded are allowed to fall onto the reverse side of the specimen, the substrate is imprinted by the knobs. The coating on the underside of the specimen is subjected to a distension or elongation determined by the radii of the individual knobs. Elongations of 0.5 to 60%, depending upon the indenter used, are possible with this type of rapid impact flexibility test and it is a very useful method of assessing the resistance of paints to minor accidents such as the impingement of stones on cars and accidental knocks to painted items.

Finally, the impact resistance of paints and particularly their resistance to chipping and to repeated surface contacts may be assessed by ball jet tests. Instruments are commercially available for these tests which are particularly valuable for testing car finishes. Typically, a stream of small balls (1 cm in diameter) or other particles is allowed to fall from a fixed height (for example, 40 cm) onto the test specimen held either horizontally or at an angle of 30° to the jet. The results of the test are then compared with those found with standards.

9.4.4 Abrasion and Scratch Resistance

Resistance to abrasion and scratching is an important characteristic in paints but, like so many other properties, it is very difficult to obtain quantitative data. In fact, abrasion, wear and scratch resistance are assessed by quite separate tests although it is difficult to define precisely the difference between abrasion/wear and scratching of surfaces.

Resistance to wear and abrasion as well as the scrubability of surfaces are assessed by means of abrasion test machines, several types of which are in widespread use. The simplest abrasion test is to allow sand or any other abrasive of controlled grain size to descend under gravity or pneumatic pressure onto the inclined test specimen from a reservoir or hopper at a specified height above the specimen. Abrasion resistance is assessed by the quantity of sand required to wear through the coating, that is, litres of abrasive per micron thickness of coating. Alternatively, sand or other abrasive particles can be projected in an air stream against the surface, that is, a miniature grit or sand-blasting apparatus is used, and abrasion/wear resistance is assessed as the reduction in thickness of the coating per specified weight of abrasive.

Abrading and scrubbing testers are perhaps the most widely used devices and these vary in both their degree of sophistication and the actual method of testing. In scrub testers, the specimen is subjected to the reciprocating action of an abrasion head which is loaded so that the abrasion head exerts a force on the coating. Single or multiple head testers having a variety of abrasion heads, such as bristle brushes, nylon, emery, etc. are available and the load on the head may be varied although the scrubbing rate is usually fixed. Alternatively, a rotational abrasive action may be employed and with such testers either the specimen is rotated under a fixed but loadable abrasing head or both the specimen and the abrasion heads rotate in opposite directions.

In these tests, abrasion/wear resistance is assessed either by the time for breakdown of the material or by measurement of the depth of wear or the volume loss of a material in a given time. Since these testers are automatic in operation, have variable loads as well as a variety of different abrasive media and can be run in the wet or the dry state, they represent a versatile and rapid method of assessing wear/abrasion resistance. Unfortunately, in many instances, it is not possible to duplicate the wear or abrasion actually experienced in service so that these tests may only yield comparative data rather than a specific indication of expected service life.

Resistance to scratching is assessed by scratch testers and these have already been discussed in Section 9.3.2. A wide variety of both manual and automatic scratch testers are available but in most scratch tests on paints, scratch resistance is determined as the maximum load on the needle or cutting device which the coating will withstand without breakdown. This test is also often referred to as a scratch hardness test.

9.4.5 Hardness

The hardness of a material, in contrast to scratch hardness referred to previously, is its resistance to indentation and this characteristic is related to the tensile strength of the material. In many cases, but not all, surface hardness can also be related to wear/abrasion resistance.

Hardness is usually determined by measuring the indentation produced by the action of an indenter of accurately known geometry, such as a hardened steel ball or a pyramidal or rhomboidal diamond, under a known weight. Both macroscopic and micro-scale hardness testers are in common use for metals and many other materials. Paint films are difficult to test in this way since they are thin and soft, although pneumatic micro-indentation hardness testers have been specially developed for paints.

The hardness of paints is often tested by pendulum or rocking testers. The Sward hardness rocker consists of a wheel constructed from two 10 cm diameter metal rings joined together. There are either bubble tubes or pendulums at the centre of the wheel to indicate the angular limit of rock of the rings. Paint hardness is assessed by determining the damping effect of the film on the rocking motion of the wheel, that is, the number of oscillations the wheel makes on the test surface, compared to that on a plate glass standard. The relation of the damping effect of the paint films to the hardness is also utilised in the König and Persoz pendulum tests, which are incorporated into various standard specifications. The basis of these tests is that a triangular pendulum resting on two tungsten carbide ball points is placed on the test surface and allowed to swing at a specified amplitude, usually between 3° and 6°. The number of swings, determined by an electronic timer, that occur on the test specimen compared with the number on a plate glass standard is related to the paint hardness. Commercial instruments are available for this test although a precise hardness measurement, comparable to that from indentation tests, is difficult to achieve.

9.5 OPTICAL PROPERTIES

In many applications, the primary function of a paint film is decorative and consequently, it is necessary for different production batches of a given paint to produce the same aesthetic effect when first applied and after exposure. Thus, the various parameters which contribute to the aesthetic appeal of a coating, notably colour, gloss and sheen, opacity and light fastness, require testing.

9.5.1 Colour

The colour of a paint (and, for that matter, of virtually all other materials) is a complex subject since the ambient light conditions, the viewing position, the pigmentation and the surface texture, as well as the inherent colour of the binder itself, all affect the colour as discerned by the viewer, who, incidentally, may or may not have defective colour perception. The latter factor, however, is not usually considered in colour measurements despite its obvious importance and, for colour assessment and comparison, trained observers with good colour vision and discrimination are usually employed.

The visible spectrum covers the wavelength range of 380 nm to 750 nm, that is, violet to red, although the eye (really the brain) exhibits the greatest response at 555 nm for all wavelengths in an equal energy spectrum from 380-750 nm.

The basis of colour measurement is that white light consists of three primary colours, red, green and blue, and all colours can be matched by admixing different amounts of these primaries. This principle is the basis of measurement in simple colourmeters or colorimeters. The test specimen is viewed through banks of different intensity filters of the three primaries and the test colour is matched with light transmitted by the filters selected. By this means, the colour can be specified with reasonable accuracy, provided a wide enough range of accurate filters is available. Certain modern instruments are still based on this principle but they incorporate standard illuminants and photodetectors and/or comparators, so that great accuracy is possible and the measurements can be converted to values on a numerical colour scale (see below).

For many years, accurate colour filters were unobtainable (and pure filters cannot be produced) so arbitrary colour scales were devised, the most important of these being the CIE (Commission International d'Eclairage) scale and the chromaticity scale. The basis of the CIE scale is that any colour can be specified in terms of the arbitrary tri-stimulus values X, Y and Z, corresponding to the three primaries red, green and blue. The system uses three standard illuminants, A, B and C. Illuminant A is a tungsten filament lamp of colour temperature 2854 K. Illuminant B is the same lamp but altered by means of correction filters to give light corresponding to that of noon sunlight. Illuminant C, the most commonly used source, is the lamp of illuminant A corrected by filters to give light representative of average daylight, that is, an overcast sky.

Colour measurements with tri-stimulus colorimeters are made by determining the percentage of reflection intensities through the tri-stimulus filters from the test surface compared to those from a perfect diffuser, usually the white reference standard, magnesium oxide. From these intensities, the X, Y and Z values are calculated and from these values, the chromaticity values or coefficients x, y and z may be calculated in turn using the relation:

$$x = \frac{X}{X+Y+Z}, \quad y = \frac{Y}{X+Y+Z} \text{ and } z = \frac{Z}{X+Y+Z}$$

where $x + y + z = 1$. (9.2)

The Y value gives the luminosity of a sample surface so that the colour and lightness can be specified by the three values Y, x and y. It is also possible, using the chromaticity coefficients, to derive the dominant wavelength or hue of light reflected from the surface by reference to a chromaticity diagram or colour triangle. This, however, is beyond the scope of this chapter and the reader is advised to consult the bibliography for a more detailed discussion of colour and colour measurement.

It should be mentioned, however, that although most colorimeters use standard illuminant C and either a photocell or the human eye in conjunction with the tri-stimulus filter systems, the viewing angles in individual instruments may vary. Commonly, the test surface is illuminated at an angle of $45°$ and is viewed perpendicularly or normally to the surface; this is known as the $45°/0°$ geometry. Other geometries, such as $45°/45°$, when the surface is illuminated at $45°$ and is viewed from the opposite side of the surface at $45°$, and

0°/diffuse, where the incident light is normal to the surface and the diffuse reflected light is measured, are also used for particular purposes.

Furthermore, whilst the CIE system is universally accepted as a means of specifying colour, it does have certain inherent weaknesses, notably non-uniformity of the colour scales with respect to the actual sample colour. Consequently, other scales, notably the Hunter, L, a_L, b_L scale, the cube root L, a, b scale and the R_d, a, b scale, have been developed. The L, a_L, b_L scale is particularly useful since it correlates closely with visually perceived colour and is mathematically related to the CIE scale. Consequently, certain tri-stimulus colorimeters are available in a choice of colour scales to facilitate instrument selection for a particular application.

9.5.2 Gloss and Sheen

Gloss, or more correctly, specular gloss, is the brightness of highlights and it is the ratio, expressed as a percentage, of the relative intensities of the incident and reflected light for the same angle of incidence/reflectance. A gloss surface has a specular reflectance greater than 50%.

Gloss is measured by glossmeters but these instruments are available with different geometries, both fixed and variable geometry devices being used. High angle instruments 20° and 45° (measured from the normal to the test surface) are used for high gloss surfaces, whilst lower angles 60°, 75°, and 85° are used for lower gloss and matt surfaces. Gloss measurements are made after the meter is set to 100% with a gloss standard and the photometer response (the meter reading) of the test surface gives the gloss as a direct percentage figure.

Gloss is an important characteristic in high quality decorative finishes but reflectance is also important. The reflectance of a surface is the ratio of the intensity of reflected light from its surface when illuminated at an angle of 45° by the CIE illuminant C and viewed perpendicularly (that is, a geometry of 45°/0° is used), to the intensity of reflected light from standard magnesium oxide. Reflectance measurements are used to assess opacity or hiding power, light fastness and weathering effects. These topics will be considered again in Sections 9.5.3 and 9.5.4.

The sheen of a surface is its gloss (or brightness) when observed at very low angles of incidence/reflection, that is, at geometries of 75° or 85°. Sheen is usually measured using low angle glossmeters.

Variable geometry glossmeters are useful because they can be utilised for virtually all surfaces. Gloss/sheen measurements are mandatory for both quality control and for assessing the effects of abrasion and environmental degradation. The latter measurements constitute rapid and non-destructive tests for assessing coating performance under a number of service and laboratory test conditions.

9.5.3 Opacity

The opacity or hiding power of a paint is the minimum thickness that is required to obliterate the substrate. Reflectance measurements (see earlier) are used to determine opacity. Opacity is determined as the contrast ratio of the paint when placed on a standard half-black and half-white tile, the two halves having the respective photoelectric reflectances of $5 \pm 1\%$ and $88 \pm 1\%$. The principle is that when the paint on a transparent substrate is placed over a white surface, any light passing through the paint is reflected by the white surface. Light transmitted by the paint when placed on a black surface, however, is absorbed. The greater the opacity, the lower is the contrast ratio. The paint is applied at a known thickness to a cellophane sheet which is then placed onto the half-black and half-white tile and the contrast ratio of the paint is determined. The contrast ratio is the reflectometer reading obtained over the black tile after the reflectometer has been set to 100% over the white tile. A variety of commercial reflectometers are available for opacity measurements but rapid opacity checks can be made with cryptometers and hiding power charts. Paints may be applied directly to hiding power charts which are available as black and white patterned sheets. The paint is applied at various film thicknesses and allowed to dry and the minimum thickness for obliteration of the substrate can be assessed directly. Furthermore, the chart may be stored as a permanent record. Cryptometers are used similarly, but with wet films, and tables are available to permit calculation of the coverage in square metres for the film thickness giving substrate obliteration.

9.5.4 Light Fastness

The light fastness of a paint is its ability to resist deterioration under the action of sunlight. Deterioration becomes evident as a colour change in the coating. Clearly, discolouration of decorative coatings is undersirable and most finishing paints are tested for light fastness.

The basic test involves preparing a painted test panel and, after covering one half of the panel with a polished metal shield, exposing it, at a distance of 25 ± 1 cm, to a carbon arc lamp. The lamp normally operates at 1300-1500 W in an enclosed glass envelope for a specified number of hours and the air temperature near the panel is maintained in the range of 38-44°C. After exposure, the uncovered portion of the test panel is compared with the shielded half of the panel as well as with a freshly prepared panel coated with the same paint but not exposed in the chamber. The coatings are compared in diffuse daylight for signs of change of colour and fading, as well as any reduction in gloss or loss of opacity. Such comparisons may be performed both by eye and by means of colorimeters, reflectometers and gloss meters in order to quantify any detectable changes.

9.6 PERMEABILITY, WEATHERING AND CORROSION PROTECTION

Paint films when applied as protective and/or decorative coatings onto a substrate must not only retain their integrity but must also provide continuous protection for the substrate throughout the life of the coating. Thus the paint itself must resist deterioration under service conditions while, ideally, presenting an impermeable barrier for the substrate against its environment. Unfortunately, all paint films incorporate defects such as pin holes, micropores and capillaries so that they always exhibit some degree of permeability to gaseous and/or liquid permeants. Furthermore, many polymers can be degraded by their environment or a film may deteriorate and exhibit a marked change in its mechanical properties due to leaching of plasticisers or other components from the films.

9.6.1 Permeability

The permeability of a paint film is a most important characteristic since permeation of environmental contaminants through to the substrate can result in loss of paint system adhesion as well as deterioration of the substrate, for example, by corrosion of metals and fungal attack and decay of timbers. Similarly, leaching of plasticisers or other film constituents can only occur if there is a continuous aqueous pathway from the environment to the inner regions of the film. These effects are more gross and generally require a degree of water absorption to become apparent.

Determinations of water absorption and desorption by a paint filmmay be made by reasonably straightforward tests, involving immersion of the specimen in water for different periods of time and weighing the film. The increased weight of the film, after removal of excess surface water, compared to the initial dry weight gives the water absorption. Weight losses on prolonged immersion indicate loss of components from the film. The desorption behaviour can be obtained by allowing the film to achieve equilibrium with water, followed by drying and periodic weighing.

Various permeability tests have been used for paints but one of the simplest is the Payne permeability cup. This is a metal cup containing a small volume of water. The opening, or aperture, of the cup is closed by clamping a film of paint over it. The cup is then weighed, placed in a dessicator and weighed at intervals. The change in weight of the cup, denoting permeation of water vapour through the film, permits calculation of the film permeability using Fick's laws of diffusion:

$$y = -D.dc/dx \qquad (9.3)$$

and
$$dc/dt = D.d^2c/dt^2 \qquad (9.4)$$

where y is the flux of permeant across unit area normal to the x direction, namely, through the film; dc/dx is the concentration gradient of the permeant at a fixed time t; dc/dt is the rate of change of permeant concentration with time at a fixed distance x; whilst D is the diffusion coefficient of the permeant in the test material.

The diffusion coefficient D of liquid water can be calculated from water absorption measurements as well as by inverting a Payne cup so that the contained water actually rests on the film. Such measurements of the diffusion coefficient assume that there is no interaction between the permeating species and the film. If such interaction does occur, as is common in many paint and other polymer systems, then the basic laws of diffusion are not directly applicable and only qualitative or comparative data may be derived.

Various other measurements of permeability, particularly of ions, can be performed using specific ion meters and radioactive tracer studies but these are outside the scope of this book.

9.6.2 Weathering Tests

Weathering of paints is the term applied to the change, usually a deterioration, that occurs from exposure in service. Various factors contribute to weathering, notably sunlight (particularly the ultra-violet component of sunlight), the ambient temperature and humidity, atmospheric pollution and geographic and climatic considerations such as rainfall. Furthermore, different parameters become important if the paint is to be used for fully or partially immersed structures in sea water or inland waters or if the paint is to be used in chemical plant. In general, however, weathering is assumed to concern only atmospheric service conditions and other tests, notably humidity and salt spray testing (see Section 9.6.3), have been devised to simulate other situations.

The best weathering/exposure trial of a paint is obviously in the proposed service environment itself but this is clearly impractical as a means of routine assessment. Service performance data on established coatings, however, must be available if the results of accelerated tests on novel formulations are to be compared with existing coatings as a basis for predicting performance.

Outdoor exposure trials using painted panels exposed under specified conditions (that is, facing south and inclined at $30°$ or $45°$ to the horizontal) and in definite locations (that is, industrial, urban, marine or rural areas) provide extremely valuable performance data. Unfortunately, such trials can last for many years and are, therefore, impractical for routine testing. Consequently, various accelerated weathering test machines have been developed commercially to provide comparative performance data and an approximately ten-fold increase in the weathering rate.

Various designs of weathering testers are used for accelerated testing and these devices range from comparatively simple constant temperature boxes containing an arc lamp, to the very sophisticated commercial testing machines. Most of these weathering machines incorporate most, if not all, of the following: programmable timers to permit simulation of cyclic night and day exposure, temperature controllers, condensation mechanisms to simulate rain and dew, facilities for controllable additions of gaseous pollutants such as ozone, sulphur dioxide and nitrogen dioxide and, finally, different light sources to permit simulation of a wide variety of exterior lighting conditions by selection of the source.

Weathering resistance may be qualitatively assessed by comparing the test specimen with the behaviour, under the same test conditions, of standard specimens having well-characterised service performances. Quantitative evaluation of weathering requires detailed examination of the changes that occur in the physical and optical properties of the test coating, combined with careful extrapolation to actual service conditions. Examination of the paint coating is often combined with inspection of the substrate for signs of deterioration.

9.6.3 Corrosion Tests

The importance of protective paints in restricting metallic corrosion has resulted in a proliferation of tests specifically designed to rapidly assess the protective ability of both novel and production coatings. In view of the great economic importance currently attached to the use of anti-corrosive paints and coatings, this section will be devoted to this aspect of testing.

The performance of paints *per se* under conditions of high humidity or complete immersion is less important than the protective effect of the paints towards metallic substrates. Consequently, various accelerated tests are performed on painted metal panels, primarily to assess the substrate corrosion but combined with an evaluation of the inherent resistance of the paint coating.

The mildest accelerated test involves exposing the painted test specimen to a high humidity atmosphere, usually 95 ± 5% rH, at 42-48°C for 48 hours or longer. Only one face of the specimen is exposed to the test environment, the other face and all edges being covered by a protective rubber, wax or lacquer coating. Since the test humidity is above the critical 70% level for steel corrosion, any breakdown or severe deterioration of the paint will be demonstrated by substrate corrosion. Variations of the standard humidity test include raising the ambient temperature in the chamber and cycling the temperature above and below the dew point so that the test specimens experience condensation.

After testing and allowing the specimen to cool to room temperature for 24 hours, the paint film is tested for adhesion to its substrate and for any deterioration in its physical properties. This procedure is usually adopted after all corrosion and other accelerated environmental test regimens, including weathering tests.

A more severe test procedure than the above is the salt spray test, which is incorporated into numerous national and international standard specifications for paints and other coatings. Unfortunately, salt spray tests are extremely aggressive, somewhat irreproducible and the results tend to correlate poorly with actual service experience. In such tests, painted panels, usually with the edges sealed with protective lacquer, wax or rubber, are exposed to a mist of 5-20% sodium chloride solution for periods of 16-96 hours and sometimes longer, usually at ambient temperatures, although elevated temperatures can be used. Various designs of salt spray cabinet are commercially available, but in all of them, there are atomisers to generate the mist and baffles to prevent direct impingement with the specimens of the droplets from the atomisers. A slightly more aggressive version of this test, but which gives better correlation with the service behaviour of electrodeposited coatings of nickel and chromium, is the acetic acid salt spray test. In this, a mist is produced of 5% sodium chloride solution, the pH of which is adjusted to 3.2-3.5 with glacial acetic acid, testing being performed at 33-37°C for 8-144 hours.

The protective efficiency of paint films is again assessed by the extent of substrate corrosion, if necessary by removal of the paint film and dissolving any corrosion products from initially weighed specimens to obtain a net weight loss as a measure of corrosion.

Salt sprays using ammonium chloride or ammonium sulphate instead of sodium chloride have been proposed for paint films. Although there is some merit to the suggested use of other salt sprays for paint testing, as yet such variations have not received universal acceptance.

A variation on the standard salt spray test is the sea water spray test. In this test, the painted specimens are sprayed with natural or synthetic sea water with a hand atomiser and then are stored under cover for a given period. The specimens are resprayed every day so that they remain wet throughout the test period. Since the specimens are exposed to the action of well-aerated sea water, this test is very severe and its aggressiveness is often increased by subjecting the test pieces to a continuous spray regimen using a pump system to recycle the sea water.

Tests for resistance to sea water and salt water under continuous immersion are also performed. In the salt water test, the specimen is half immersed in a 3.5% solution of sodium chloride which is exposed to the atmosphere. Thus, the painted metal is subjected to the action of well aerated saline solution at the water line but poorly aerated water towards the bottom of the container. In contrast, for sea water resistance tests, the specimen is fully immersed in natural or artificial sea water which is aerated by means of a bubbler. This test is also somewhat aggressive. After both types of testing, the paint film is assessed for damage and the metal, after removal of the paint by means of a suitable solvent, is also inspected for any corrosion.

Several other corrosion tests have been developed for metallic coatings but most of these are rarely used for paints. Furthermore, many electrochemical techniques have been developed to measure corrosion rates as well as to assess corrosion susceptibilities. These techniques, however, are very specialised in nature and despite their obvious usefulness and importance in the general corrosion field, they are not widely used in the paint field as yet. They are discussed in the books cited in the bibliography, to which reference should be made for further information.

In this chapter, the multiplicity of tests for paints and paint coatings have been discussed. It is clear that the field of paint testing is both wide and complex but it is a necessary aspect of paint technology if new and improved coatings are to be developed now and in the future.

9.7 BIBLIOGRAPHY

Instrumentation in Metal Finishing, J.A. von Fraunhofer, Elek Science, London (1975).

Basic Metal Finishing, J.A. von Fraunhofer, Elek Science, London (1976).

Protective Paint Coatings for Metals, J.A. von Fraunhofer and J. Boxall, Portcullis Press, Redhill, Surrey (1976).

Physical and Chemical Examination of Paints, Varnishes, Lacquers and Colours, H.A. Gardner, ASTM Special Technical Publication 500, Philadelphia, Pa. (1968).

Paint Technology Manual, Part 5; The Testing of Paints, D.S. Newton (ed.), Chapman and Hall, London (1965).

Concise Corrosion Science, J.A. von Fraunhofer, Portcullis Press, Redhill, Surrey (1975).

Handbook of Corrosion Testing and Evaluation, W.H. Ailor (ed.), J.Wiley, New York (1971).

British Standard 3900: Methods of Test for Paint, Groups A to G.

ISO Recommendations: TC35; R1516, R1518, R1519, R1520.

Index

Abel tester 172-173
Abrasion cleaning 136-137
  resistance, 46, 47, 48, 89,
   113, 114, 116, 140-141,
   145, 148, 149, 163, 187-189
Accelerated corrosion tests
  see Corrosion tests
  weathering tests 196-197
Acetone 91, 94, 96
Acid Pickling 136, 137
  resistance 36, 37, 39, 40,
   50, 56, 59, 62, 72, 73, 77,
   78, 79, 81, 82, 163, 164
  wash 137
Acrylic resin 39-40, 42, 43,
   92, 97, 115, 141, 150, 151,
   158, 159, 179
Acrylonitrile 40
Addition polymerisation 5-7,
   15, 29
Additive 58-59, 81-83, 84-107,
   169, 174
Adduct 42, 47, 48
Adhesion 3, 36, 38, 39, 43,
   44, 49, 51, 70, 137, 140-141,
   142, 146, 147, 166, 181-184,
   194, 197
Ageing see Durability
Air compressor 125-126

Air drying 30, 33, 38, 43,
   46, 52, 77, 78, 95, 97,
   131, 142, 143, 144, 152,
   156, 157, 158, 164, 167
Airless spray 125, 128,
   129-130, 145, 151
Alcohols 31, 44, 45, 51,
   56, 76, 91, 93, 94, 96-97
Alcoholysis process 32
Aliphatic solvents 33, 51,
   76, 78, 93, 94, 96
Aldehyde 35, 48, 51
Alkali resistance 34, 36,
   37, 39, 40, 44, 50, 56,
   59, 62, 66, 67, 70, 72,
   73, 77, 78, 79, 81, 148,
   149, 151, 152, 164
Alkaline cleaning 138
Alkyd resin 2, 30, 31-35,
   36, 43, 44, 47, 49, 50,
   85, 86, 92, 95, 97, 98,
   106, 109, 110, 111, 112,
   115, 140, 145, 146, 148,
   154, 155, 156, 157, 158,
   159, 163, 164
  manufacture of 31-32, 34
  modified 34, 46, 55, 85,
   95, 97, 106, 143, 144,
   154, 156-157, 158, 164,
   165

Aluminium driers 97
  pigment 65, 74-75, 83, 158, 159, 165
  substrate 70, 147, 148
Amine 35, 42, 44, 45, 48, 105
Amino resin 33, 34, 35-36, 42, 43, 44, 97, 115, 143, 154, 156-157, 165
Anatase *see* Titanium dioxide
Anodic polarisation 139
  reaction 138
Anti-corrosion paint 68, 70, 71, 72
  see also Corrosion resistance
Anti-fouling agent 101-102
Anti-oxidant 6, 63
Anti-settling agent 106-107, 131
Antimony oxide 64
Application of paint 123-134
  rate 110
Aqueous dispersion *see* Emulsion
Aromatic solvents 33, 37, 49, 51, 76, 78, 93, 94, 164
Arylamide red 77
Asbestos-cement 148
Atactic polymer 9, 11, 12
Atmosphere *see* Location and Weathering
Attrition mill 119, 120-121
Auto-oxidation 31
Auxochrome 24
Azo pigment 75, 76-79

B

Bacteria 99, 102
Bakelite 7, 179

Baking finish *see* Stoving cure
Ball mill 36, 74, 119, 120
Barrier film 75, 139, 142
Barium metaborate 100
Barytes 81-82, 145, 160
Basic lead carbonate *see* White lead
  silicochromate 68
  sulphate 65
Basic pigment 64, 65, 70, 71
Bend test 181-182, 185
Bentonite clay 105
Benzidine yellow 78-79
Benzotriazole 157, 158
Benzoyl peroxide 5, 6
Beta ray meter 180-181
Binder 2, 4, 19, 27, 29, 30-56, 84, 97, 98, 101, 108, 109, 112, 113, 119-121, 135-161, 165, 166, 169, 172
Biocide 98-99, 101, 150, 151, 159
Bisphenol-A 41, 42
Bituminous paint 1, 55-56
Black iron oxide 74
Black pigments 74, 80-81, 168
Blanc fixe 81-82
Blast cleaning 136, 137, 138, 147
Bleeding 76, 79, 80, 164
Blistering, resistance to, 114
Blocked adduct 47
Bloom 98
Blue pigments 72-73, 80
Brass 147, 148
Brittleness 48, 50, 64, 84, 166, 167, 168
Brush application 94, 95, 96, 106, 123-124, 125, 128, 144, 145, 151, 152, 157, 158, 170

Bulk polymerisation 36
Butanol 36, 97, 146, 155, 156
Butyl acetate 91, 94, 96
   alcohol see Butanol
   stearate 89

C

c.p.v.c. 113, 114
CIE system 191-192
Cadmium reds 67, 165
   yellow 71, 165
Calcium carbonate see Whiting
   organic salts 98
   plumbate 71-72
   salts 98, 107, 111, 160
Carbon arc lamp 194
   black 16, 80-81, 165
Carboxylic acid 31-32, 42
   see also Fatty acid and
   Polycarboxylic acid
Castor oil see Dehydrated
   castor oil
Catalyst 5, 34, 36, 39, 41,
   42, 55, 154, 155, 156, 167
Cathodic polarisation 138
   reaction 139
Cellosolve 96
Cellulose 33, 50-51, 121,
   150, 151, 154, 159, 173
Cellulose ether 96, 106, 150
   nitrate see Nitrocellulose
   solutions, regulations for,
   173
Cement colourants 66, 72,
   73, 79
   render 148
Cementitious substrate see
   Substrate
   surface preparation 152-153
Chalk see Whiting
Chalking 62, 63, 64, 65, 166
Charpy test 186

Chemical resistance 24, 34,
   36, 37, 38, 40, 43, 44, 46,
   47, 48, 50, 51, 52, 55, 56,
   59, 61, 62, 66, 67, 72, 73,
   74, 77, 78, 79, 80, 81, 89,
   101, 113, 140-141, 144, 145,
   146, 147, 148, 149, 151,
   163-164
China clay 82
Chipboard 154
Chloride ion 70
Chlorinated paraffin 89, 144,
   145, 164
   rubber 34, 49-50, 84, 85,
   89, 92, 93, 95, 112, 115,
   141, 144-145, 150, 152,
   163, 164, 167
   solvent 37, 164
Chromate pretreatment 137,
   147
Chromic acid 70
Chromium oxide 72
Chromophore 23, 24
Classification of paints 1-2
Clay 105
Climate see Weathering
Coal tar epoxy resin 43,
   55, 81
Coalescing solvent 150
Cobalt drier 97, 98, 111,
   158, 159, 160
Cohesive strength see also
   Mechanical properties
Cold-cure resin 42, 47,
   48, 56, 179
Colloidal silica 155
Colorimeter see Colour
   measurement
Colour 2, 3, 20-22, 58, 59,
   61, 66, 85, 114, 121, 122,
   143, 190-192
   additive mixing 22
   fastness 23-24, 43, 59,
   61, 63, 65, 67, 68, 70,
   72, 73, 76, 77, 78, 80,
   81, 88, 165, 166, 193-194

matching 121-122, 190-192
measurement 22, 122,
  190-192, 194
stability 63, 65, 67, 76,
  85, 101, 164, 192,
  193-194, *see also*
  Yellowing
subtractive mixing 22
Compatibility of resins 93
Composition graph 116, 117, 118
Concrete 124, 148, 166
Condensation polymerisation
  7-9, 29, 31, 32, 35, 41, 51,
  53, 54, 103
Condensation conditions,
  resistance to 163, 166-167
Conductivity, of paints 130
Conjugation 33, 34
Contamination, surface 138, 147
Contrast ratio 27, 28, *see also* Opacity
Convertible coating 29-30, 40, 89
Copolymerisation 4, 7, 11, 14, 34, 38, 40, 42, 43, 55, 84, 150
Copper 102, 147, 148
Corrosion 102, 135, 138-139, 144, 147, 166, 194, 197
  product 135-136, 138
  resistance 2, 34, 46, 50, 55, 56, 65, 68, 70, 71, 75, 83, 140-141, 144, 178, 193, 197-199
  testing 197-199
Cost 113, 116, 117, 124, 128
Coumarone resin 31
Covering power 27
Cracking of films 168
Crepe rubber 49
Critical pigment volume
  concentration (c.p.v.c.)
  113, 114

Cross-linkage 8, 9, 12, 15, 33-34, 36, 40, 41, 42, 43, 46, 47, 48, 52, 54, 97, 164, 165, 167
Cryptometer 193
Crystallinity 4, 9, 11, 15, 39
Cupping test 181, 182, 185-186
Curing of paint *see* Drying
Curtain coating 133, 156, 157

D

D.C.O. 30
Decorative finish 33, 66, 67, 68, 69, 75, 77, 78, 79, 80, 81, 83, 95, 109, 148, 162, 194, *see also* Finishing paints
Degreasing 95, 138, 147
Dehydrated castor oil 30
Delamination of oxides 136, 137
Density *see* Specific gravity
Depth gauge 175-176, 179
Deterioration *see* Durability
Dibutyl phthalate 88, 155
Dichlorfluamide 101
Differential aeration cell 102
Diluent 89, 93, 95, 96
Dimensional change 178
Dimethyl ketone *see* Acetone
Dioctyl phthalate 88
Dip coating 123, 130-131, 133, 158, 170
Dirt retention 151
Discolouration *see* Colour stability and Yellowing
Dispersion of pigments *see* Pigment
Dithiocarbamates 107
Double bond 4, 6, 9, 15, 34

Driers   2, 30, 34, 43, 46,
   47, 97-98
Drying of paint   29, 33-34,
   38, 40, 44, 89, 93, 96,
   98, 114, 169, 170, 173
Drying oil   30, 33, 43, 55,
   97
   alkyds   32, 33, 158, see
   also Alkyds
   rate   34, 97, 101, 105,
   145
   time   89, 93-101, 114,
   124, 144, 158, 163, 167,
   173, 176-177
Ductility   114, 181, 184,
   185-186
Durability   2, 16, 43, 46,
   49, 52, 55, 56, 61, 81,
   83, 93, 98, 99, 100, 109,
   113, 114, 135, 136, 150,
   151, 152, 154, 157, 162-168,
   194, see also Weathering
Dye   see Pigment

E

Eddy current gauge   178, 179,
   180
Edge runner   119, 120
Edwards test   185
Eggshell finish   82
Elasticity   65
Elastomer   54
Electrodeposition of paint
   131-132, 170
Electrostatic spraying   123, 130
Elongation   see Ductility
Embrittlement   98, 167
Emulsion systems   40, 62, 78,
   80, 82, 88, 98, 99, 101,
   105, 106, 109, 110, 115,
   116, 121, 124, 149, 150-152,
   158, 163, 167
Emulsion polymerisation   36-37
Epichlorhydrin   41, 42

Epoxide grouping   40, 42
Epoxy ester   42, 43
   resin   40-44, 55, 92, 96,
   133, 141, 144, 146, 147,
   153, 163, 167, 179
Erichsen tester   182
Erosion   109
Ester linkage   34, 43
   solvent   78, 91, 92, 93,
   95-96, 164
Esterification   31, 32, 43,
   51
Etch primer   39, 70, 97,
   142-143, 144-145, 147,
   148
Ethers   37
Ethyl acetate   91, 94,
   95-96
   alcohol   96, 159
   hydroxy ethyl cellulose
   106
Ethylene glycol monoethyl
   ether   96
   oxide condensates   103
Evaporation rate   89,
   93-94, 95, 96
Exposure   see Weathering
Extender   2, 18, 25, 27,
   81-83, 108, 109, 112,
   113, 116, 119, 145, 146,
   153
   function in paints   2, 25,
   58-59, 145, 153
Extensibility   see Ductility
Exterior finish   109, 115,
   144, 157-160, 162-168,
   196-197

F

Factory finishing   123, 125,
   145, 156, 157, 158, see
   also Industrial paints
Fatty acid   30, 31, 32, 33,
   34, 42, 44, 46, 103

Fatty acid process 32
Ferrous substrate 65, 67, 70, 71, 75, 102, 135-147, 166, 179
Fick's laws 195
Filler 2, 66, 74, 81, 82, 119, 153, 154, 160
Filling out, of paint, 120
Film degradation 62
  durability *see* Durability
  formation *see* Drying
  thickness 25, 27, 95, 105, 110-111, 128, 133, 139, 142, 151, 152, 154, 163, 175-176, 178-181, 193
Fineness of grind 174-175
Finishing coat 3, 33, 56, 65, 68, 81, 82, 83, 106, 109, 110, 120, 130, 131, 138, 139, 143, 145, 146, 148, 149, 152, 154, 158, 159, 160
Fire retardant paint 64
Flaking 166
Flame cleaning 137, 138
Flammability 88, 170, 172-173
Flashpoint 170, 172-173
Flat finish 18, 24, 82, 115
Flatting agent 82, 109, 154
Flexibility 11, 15, 39, 40, 43, 47, 49, 51, 83, 84, 85, 86, 87, 113, 140-141, 185-186, 187
Floating (flooding) 72
Floor paint 116
Flow coating 131, 132-133, 170
  cups 170
  properties 59, 65, 68, 82, 93, 94, 95, 96, 104-106, 124, 170-172 173-174, *see also* Viscosity and Rheological properties

Flux-meter 180
Formaldehyde 35, 36, 51, 99, 156
Formulation of paints *see* Paint
Fouling 101-102
Fungal growth 64, 99-100, 160
Fungicide 2, 64, 99-101
Furniture lacquers 97, 154

G

Galvanised steel 71, 147
Garmson gauge 175
Gel 104, 105, 106
Gibbs free energy 90
Glass transition temperature *see* Tg
Gloss 3, 18, 24, 55, 56, 59, 63, 65, 82, 89, 109, 110, 113, 114, 115, 146, 154, 157, 158, 160, 163, 167, 174, 192-193, 194
Gloss meter 192-193, 194
Glycerol 31, 52
Glycol ether solvents 93, 96
Graphical approach in paint formulation 116-117
Gravity feed gun 127, 152
Green pigments 72, 73, 80

H

Haematite 65-66
Hansa yellow 78
Hard dry time 177, *see also* Drying time
Hardboard 154
Hardwoods 153, 157
Hardness 34, 47, 48, 89, 146, 189, *see also* Mechanical properties
Heat resistance *see* Temperature

Hegman gauge 175
Helio reds 75-77
Hiding power *see* Opacity
 charts 193
High build systems 139, 145,
 152, 155
Homopolymer 7, 38
Hot spray technique 128-129,
 145
Hue 21
Humidity 100, 167, 197
Hunter colour scale 192
Hydrocarbon resins 152
 solvent 37, 55, 91, 92,
 93, 94, 95, 96
Hydrogen bonding 91, 92, 93
Hydrolysis 101
Hydroxy-cured polyurethanes
 48-49

I

Impact resistance 140-141,
 186-187, *see also*
 Mechanical properties
In-can stability *see*
 Storage
Indene resin 31
Indenter tests 186-187,
 189
Industrial alcohol *see*
 Ethyl alcohol
 atmosphere 68, 162-163
 paint 66, 68, 80, 81,
 95, *see also* Factory
 finishing
Intra-red radiation 29
Inhibitive pigment 138, 139
Inhibition, of corrosion, 68,
 70, 138, 139
 of polymerisation 6
Initiator *see* Catalyst
Inorganic pigments 23, 59,
 61-74
 silicate 56, 144, 165

Interior finish 109, 115,
 116, 154-157, 163
Iron *see* Ferrous substrate
 oxides 23, 65-66, 71, 74,
 143, 165
Irradiation 15
Isocyanate 34, 44, 47,
 48, *see also* Polyurethane
 resins
Isomerised rubber 74
Isoprene 29
Isotactic polymer 9, 11, 12,
 37
Izod test 186

K

Kaolin 82
Kauri resin 31
Ketone solvents 37, 78, 91,
 92, 93, 95-96, 155, 164
Knotting 159
König tester 189

L

Lacquers *see* Varnish
Lapis Lazuli 73
Leaching 86, 102, 194, 195
Lead chromates 68-69, 72, 77,
 78, 79
 chrome green 72, 73
 driers 97-98
 pigments 2, 18, 64-65,
 71, 75, 111, 137, 143,
 158, 159, 160
 silicochromate 68
 soaps 65, 66, 71, 75
 substrate 147
 sulphate 65
Lemon chrome 69
Levelling 105, 106, 124,
 151
Lewis acid 5

Light fastness *see* Colour fastness
Linear polymer 8, 35, 37, 40, 44, 54
Linoleate driers 98
Linseed oil 30, 32, 34, 52, 55, 56, 67, 86, 92, 98
Location 162-165 *see also* Weathering

M

MDI 44, 48
Magnesium 97
Magnetic gauge 178-180
Maleic acid 38
Mandrel test 181-182, 185, 186, 187
Manganese driers 97-98
Manual cleaning 136, 147, 159
Mar resistance 140-141
Marine coating 65, 101-102, 162
Masonry *see* Cementitious substrates
Mastics 2, 67, 119
Matt finish *see* Flat finish
Mechanical cleaning 136-137, 147, 159
  deformation tests 181-182
  properties of coatings 11, 15, 38, 39, 43, 44, 46, 49, 50, 52, 54, 55, 56, 59, 64, 65, 67, 71, 81, 82, 83, 84-89, 113, 114, 140-141, 165, 178, 184-190
  thumb test 177
Melamine 33, 35, 112, 146
Mer *see* Monomer
Mercury compounds 99, 100, 102, *see also* Organomercury
Metallic pigment 74-75, 115
Methylated spirit *see* Ethyl alcohol
Methylolurea 35

Methyl isobutyl ketone 96
Methyl methacrylate *see* Acrylic
Mica 83, 145
Micaceous iron oxide 65
Microsectioning 179
Mid-chrome 69
Mill scale 136, 137, 138, 147
Modified alkyd resin *see* Alkyd resin
Moisture cure 47, 48
  resistance *see* Water resistance
Monomer 4, 7, 29
Mould growth *see* Fungal growth

N

Naphthenate driers 97, 98, 111, 158, 159, 160
Natural clay 105
  resins 31, 95, 97, 155
  weathering test *see* Weathering
Network polymer 15, 36, 51
Newtonian fluids 171
Nitrocellulose 33, 34, 50-51, 84, 86, 88, 89, 92, 95, 96, 97, 112, 115, 131, 154, 155, 156
Non-convertible coating 29-30, 38, 40, 51, 89, 90, 93, 145, 164, 167
Non-drying alkyds 33
  oils 30-31, 33, 43, 44, 97
Non-ferrous substrate 147-148, 179
Non-Newtonian fluids 171
Non-volatile content *see* Solids content
Novolac resin *see* Phenolic resin

208

O

Odour 89, 95
Oil absorption 17, 18, 19
  length 31, 33, 43-44,
    46, 52, 95, 111, 144,
    155, 158
  modified resin 46, 49,
    52, 92, 95, 140, 141,
    143, 146, 155, 158,
    159, 160, 164
  resistance to 39, 164
Oils see Vegetable oils
Oleoresinous binder 30-31,
  33, 43, 65, 95, 97, 143,
  163, 164, 165
  varnish see Varnish
On-site finishing 123,
  125, 136, 157
Opacity 1, 18, 19, 20,
  24-28, 58, 59, 61, 62,
  68, 72, 76, 78, 79, 80,
  81, 153, 174, 175, 192,
  193, 194
Orange chrome 69
Organic, resistance to acid
  66
  peroxide 154, 156
  pigment 23-24, 59, 75-81,
    164, 165
  silicates 56, 165
Organomercury compound 99,
  100-101
Organosol 37-38, 88, 165
Organotin 54, 101, 102
Ostwald viscometer 171
Oxidative polymerisation 30,
  33, 130, 167
Oxide layer 135, 136, 137
Oxidising agent 36, 97

P

PMMA see Acrylic
p.v.c. see Pigment volume
  concentration

PVA see Polyvinyl acetate
PVC see Polyvinyl chloride
Paint application 42, 104,
  110, 123-133, 140, 141
  curing see Drying
  formulation 103-122,
    138-160, 162, 164, 169
  function of 1, 2
  inspection gauge 181
  manufacture 16, 18, 84,
    102, 103, 106, 110,
    119-122, 174
  testing 169-199
  viscosity see Viscosity
Paris white 82
Particle size of resins 36,
  37
  jet test 187
Payne cup 195
Peel test 181, 182-183
Penetration depth gauge
  see Depth gauge
Pensky Martens tester
  172-173
Pendulum testers 186, 189
Percentage weight 108, 110
Permeability 39, 50, 59,
  83, 87, 100, 102, 114,
  194-195, see also Water
  permeability
Permanent red 25, 76-77
  FRLL see Arylamide red
Peroxide catalyst 5, 34,
  36, 154, 155, 156
Persoz tester 189
Perspex see Acrylic resin
Phenol 44, 45, 47, 51-52
  biocides 99
Phenolic resin 7, 42, 43,
  51-52, 86, 141, 142, 147,
  158
Phosphate treatment 137,
  138
Photochemical effects
  62-63, 65, 66, 166
Phthalic anhydride 31
Phthalocyanine blue 80

209

Pickling 136, 137
Pigment 3, 18, 56, 67, 70,
  71, 74, 82, 107, 108, 109,
  110, 111, 112, 113, 115,
  116, 119-122, 132, 138,
  139, 142, 143, 150-151,
  152, 164, 169, 178
 dispersing agents
  102-103, 107
 dispersion 16-20, 81,
  102-103, 106, 119-122,
  174-175
 function in paints 2, 16,
  25, 58-59, 168
 Green B 79
 particle size 16-20, 27,
  36, 62, 82, 102-103,
  107, 174-175
 settlement see Settlement
 texture 18
 volume concentration 110,
  112-113, 114, 143, 145
Pigment-binder ratio 109,
  112, 113, 115, 152, 153
Pitch epoxy coating see
  Coal tar epoxy
Plaster, gypsum 148, 149,
  152-153
Plasticiser 4, 33, 37, 50,
  51, 84-89, 167, 194
Plastisol 37-38, 88, 165
Plywood 154, 157
Polarisation 65, 71, 139
Pollution 162-168, 196,
  see also Chemical resistance,
  Durability and Weathering
Polyamide 42, 153
Polyamine 42
Polycarboxylic acid 32,
  38, 103
Polyester 133, 154-156
Polyhydric alcohol 31, 32,
  44, 47, 48
Polymer 2, 4, 9-15, 29-55,
  85, 90-91, 110, 111, 166
Polymerisation reaction 4-9
Polystyrene 144

Polyurethane resin 34,
  44-49, 55, 92, 95, 96,
  133, 135, 140, 148,158,
  163, 164, 165, 167
Polyurethane-alkyd see
  Urethane-alkyd
Polyvinyl acetate 7, 10,
  11, 37-38, 88, 150, 151
 alcohol 11
 butyral 39, 92, 97
 chloride 7, 10, 36-38,
  84, 86, 88, 167
 copolymers 38, 39, 84,
  93, 150
Polyvinylidene chloride 7,
  11, 13, 39
Porosity, substrate 149,
  152, 166, 194
Pot life 42, 49, 56
Prepolymer 47
Pressure feed gun 127-128
Pretreatment 137, 147 see
  also Etch primer
Primers 3, 18, 56, 74, 81,
  82, 107, 109, 110, 115,
  120, 138-148, 149, 163,
  for metals 66, 67, 70,
   71, 75, 131, 132, 136,
   142-144, 145, 146,
   147-148
 for timber 65, 158
Primrose chrome 69
Promoter see Catalyst
Protective coating 68, 71,
  74, 83, 95, 138-139,
  140-141
 pigment see Corrosion
  resistance
Prussian blue 24, 59,
  72-73
Pug mill 119
Pull-off test 183-184
Putty 2, 119
Pycnometer 172

Q

Quality, of paint 109, 110, 122, 169-200
Quaternary ammonium compounds, 103

R

Radiation cure 15, 29
Red iron oxide 65-66, 143, see also Iron oxide
lead 2, 66-67 68
pigments 65-68, 75-77
Reducing power of pigments 20
Redwood viscometer 171
Reflectance 27, 192, 193
Reflectometer 193, 194
Refractive index 25, 26, 28, see also Opacity
Relative humidity, (rH) 100, 167, 197
Resin see Polymer
exudation 158
Resistance, polarisation 139
Resol 51-52 see also Phenolic resin
Rheological properties 19, 65, 68, 70, 71, 82, 93, 94, 95, 96, 104-106, 119-121, 124, 155, see also Flow and Viscosity
Rocker hardness tester 189
Roller coating 123, 124-125, 133, 151, 152
mill 119-120
Rosin 31, 52, 73, 98
Rosinate drier 98
Rust see Corrosion product
Rutile see Titanium dioxide

S

Sacrificial protection 144

Sagging 94, 105, see also Flow
Sand-cement render see Cement render
Sand mill 119, 121
Sand textured paint 151-152
Salt spray test 197-198
Saponification 64, 65, 66, 148, 164
Scarlet chrome 69
Scattering power 24, 25, 26
Scratch resistance 89, 183, 187-188, see also Mechanical properties
tester 183, 188, 189
Scrub resistance 116, 188
Sea water 43, 56, 196, 197-199
Sealer 2, 3, 104, 149, 150, 152, 154, 157
Secondary bonds see Valence bonds
Self-cleaning 63
Self-cure see Etch primer
Semi-drying oil 30-31, 32, 33, 43, 44, 97, 111
alkyd 32, 33
Service environment 1, 135, 162-168, 196-197
Setting of paint see Drying
Settlement 18-19, 73, 75, 81, 82, 106-107, 123, 131
Shear rate 104, 105, 119-121, 170-172
Sheen 192-193
Shellac 159
Shot blasting see Blast cleaning
Silica 73, 155
Silicate see Inorganic silicate
Silicate clay 105
Silicone resins 34, 52-55, 165
Soap formation see Saponification
Sodium carboxy methyl cellulose 106

Softwoods 153, 154, 158, 163
Solids content 109-110, 115, 128, 152, 155
Solubility parameters 90-93
Solution polymerisation, 36-37, 38
Solvents 2, 11, 33, 36, 37, 49, 51, 55, 58, 76, 77, 78, 79, 80, 81, 85, 86, 89-97, 105, 109, 110, 123, 128, 129, 153, 154, 164, 167, 171, 172-173, *see also* Individual headings
  cleaning 95, 138
  power 90-93, 95, 96
  resistance 9, 11, 15, 55, 76, 78, 79, 80, 81, 93, 95, 164
  solubility 11, 25, 33, 75, 88, 90-93, 103
Soya bean oil 30, 32
  lecithin 106
Specific gravity 17, 18, 19, 110, 111, 112, 172, 178
Spray fog 126, 129
  gun 125, 126, 127, 128, 129, 152, 156, 157
  painting 38, 94, 95, 98, 123, 125-130, 133, 139, 144, 145, 146, 147, 152, 154, 156, 157, 170
Stability of paint *see* Storage
Stain finish 66, 104
Staining power *see* Tint strength
Stearic acid 74
Steel *see* Ferrous substrate
Stokes' law 18, 19
Stopper 2, 81
Storage of paint 18, 49, 67, 68, 71, 82, 98, 104, 105, 106, 115, 142, 172-173

Stoving cure 33, 38, 44, 46, 47, 48, 52, 55, 67, 76, 77, 78, 79, 95, 130, 131, 132, 133, 142, 143, 146, 147, 156, 157, 158, 165
Stress, internal 174
Styrenated acrylic 150
  alkyd 34, 154
Styrene 34, 40, 154, 156
Substrate, cementitious 1, 68, 70, 73, 77, 79, 124, 148-153, 166
  metal 1, 3, 70, 75, 133, 135-148, 179, 194, 197-199
  timber 1, 3, 65, 66, 133, 153-161, 166, 194
Subtractive colour mixing 22
Suction feed gun 127
Sulphate reducing bacteria 102
Sulphonated oils 103
Sulphur oxides 68, 135, 163, 196
Surface active agent 103, 106, 150
  contamination *see* Contamination
  drying time *see* Drying time
  preparation 136-138, 147, 152-153, 159-160, *see also* Substrate
  tension 174
Suspension polymerisation 36-37
Sward rocker 189
Syndiotactic polymer 9, 11, 12

T

TDI 44, 46, 47, 48, 92
T_g 15, 36, 40, 42, 85, 86

TiO$_2$ see Titanium dioxide
Tack-free time 177
Talc 83, 145, 160
Tar see Bituminous paint
Temperature, effect on films
    100, 150, 151, 165, 167-168,
    184
    resistance 55, 62, 66, 67,
    72, 73, 74, 76, 77, 78,
    79, 80, 88, 164-165,
    167-168
Tensile strength 87, 88,
    114, 184-185, 189, see
    also Mechanical properties
    testing 182, 184, 185
Thermoplastic resin 9, 15,
    37, 38, 40
Thermosetting resin 8, 15,
    40, 43
Thickening of paint 104-106,
    116, 121, 150, see also
    Thixotropy
Thickness of coatings see
    Film thickness
Thinner 109, 171, see
    also Solvent
Thixotropic agent 105-106,
    109
Thixotropy 94, 104-106,
    109
Timber see Substrates
    pretreatment 159-160
Tin compounds see Organotin
    compounds
Tinplate 147
Tinting strength 18, 19,
    20, 61, 66, 68, 73, 74,
    80, 81
Titanium dioxide 16, 25,
    61-63, 64, 111, 145, 146,
    151, 152, 153, 157, 159,
    160, 165, 166
Toluene 91, 93, 95
Toluidine red 18, 75-77
Toner 21, 79
Top coat see Finishing
    coat

Torsional adhesion test
    184
Toxicity 49, 63, 65, 66,
    68, 77, 78, 80, 85, 89,
    98, 99, 101, 140-141
Toxin 96, 99, 102
Tributyl tin oxide 101
Trichloroethyl phosphate
    (TCEP) 88
Tricresyl phosphate 88
Triphenyl phosphate 88
Tri-stimulus colorimeter
    see Colour measurement
Tung oil 30, 32, 34, 52,
    55, 56, 158, 159
Two-pack formulation 42

U

Ultramarine blue 59, 73
Ultra-violet light 29, 66,
    100, 101, 157, 165-166,
    193-194, 196
Undercoat 3, 18, 68, 74,
    81, 82, 107, 109, 115,
    120, 139, 145, 158, 160,
    163
Underfilm corrosion 137,
    142, 166, 181
Unsaturation 4, 30, 97
Uralkyd resin 34, 46, 47,
    115, 158, see also Alkyd,
    modified and Polyurethane
    resins
Urea 44, 45, 112, 156
Urea-formaldehyde resin 8,
    35-36, 156-157
Urethane see Polyurethane

V

Valence bonds 9, 15, 90,
    165-166
van der Waals forces 9,
    11, 15

Varnish 3, 31, 43, 52, 86, 95, 97, 104, 143, 148, 157, 158, 163, 164
Vegetable oil 30, 42, 44, 46, 49, 81, 95, 97, 106, 112, 165
Vehicle 29, 31
Vinyl resins 4, 5, 7, 9-13, 34, 36-39, 92, 93, 95, 96, 112, 115, 131, 141, 144, 148, 150, 152, 163, 164, 167
Vinylidene resins *see* Polyvinylidene resin
Viscometer 171
Viscosity 49, 67, 89, 93, 98, 104-106, 115, 119, 120, 121, 123, 125, 128, 130, 131, 133, 154, 170-172

W

Wash primer *see* Etch primer
Wash resistance 113
Water absorption 48, 194-195
  permeability 83, 102, 194-195
  resistance to 34, 39, 43, 46, 49, 50, 55, 74, 83, 113, 140-141, 142, 148, 149, 150, 152, 166-167, 194-195, 196
  solubility of pigments 68, 70, 80
Water-based paint *see* Emulsion paint
Wear resistance *see* Abrasion
Weathering machines 194, 196-197
  resistance 40, 55, 56, 62-63, 64, 72, 82, 113, 137, 138, 139, 140-141, 143, 144, 148, 149, 162-163, 165-168
  tests 196-197
Weight per volume 110-113, 172
Wet blasting 137,
  edge time 124, 176-177
  film thickness 111, 175-176
Wetting of pigments 19, 102-103, 119-121
White lead 64-65
  pigments 19, 25, 61-65, 168
  spirit 91, 94, 95, 111, 133, 145, 158, 159, 160
Whiting 82, 107, 159, 160
Wire brushing 136, 137, 147

X

Xylene 91, 93, 95, 144, 145, 146, 156

Y

Yellow iron oxide 71,
  see also Iron oxides
  pigments 68-72, 78-79
Yellowing 34, 46, 88, 106, 165

Z

Zinc chromates 69-70, 142, 143, 147, 148
  oxide 63-64, 100
  pigment 63, 75, 144
  substrate 147
  tetroxychromate 70
Zinc-rich primer 56, 75, 144
Zirconium drier 97, 98

214